The International Library of Sociology

THE JOURNEY TO WORK

Founded by KARL MANNHEIM

The International Library of Sociology

THE SOCIOLOGY OF WORK AND ORGANIZATION
In 18 Volumes

I	Apprenticeship	*Liepmann*
II	Industrial Disputes	*Eldridge*
III	Industrial Injuries Insurance	*Young*
IV	The Journey to Work	*Liepmann*
V	The Lorry Driver	*Hollowell*
VI	Military Organization and Society	*Andrzejewski*
VII	Mobility in the Labour Market	*Jeffreys*
VIII	Organization and Bureaucracy	*Mouzelis*
IX	Planned Organizational Change	*Jones*
X	Private Corporations and their Control - Part One	*Levy*
XI	Private Corporations and their Control - Part Two	*Levy*
XII	The Qualifying Associations	*Millerson*
XIII	Recruitment to Skilled Trades	*Williams*
XIV	Retail Trade Associations	*Levy*
XV	The Shops of Britain	*Levy*
XVI	Technological Growth and Social Change	*Hetzler*
XVII	Work and Leisure	*Anderson*
XVIII	Workers, Unions and the State	*Wootton*

THE JOURNEY TO WORK

Its Significance for Industrial and Community Life

by

KATE K. LIEPMANN

With a Foreword by
A. M. CARR-SAUNDERS

Routledge
Taylor & Francis Group
LONDON AND NEW YORK

First published in 1944 by
Routledge, Trench, Trubner and Co., Ltd

Reprinted in 1998 by
Routledge
2 Park Square, Milton Park, Abingdon, Oxon, OX14 4RN

Simultaneously published in the USA and Canada by Routledge

711 Third Avenue, New York, NY 10017

Reprinted 2000

Transferred to Digital Printing 2007

Routledge is an imprint of the Taylor & Francis Group

Transferred to digital print 2013

British Library Cataloguing in Publication Data
A CIP catalogue record for this book
is available from the British Library

The Journey to Work

ISBN 978-0-415-17678-1 (hbk)
ISBN 978-0-415-86257-8 (pbk)

Publisher's Note
The publisher has gone to great lengths to ensure the quality of this reprint but points out that some imperfections in the original may be apparent

FOREWORD

THERE are only a few inventions of which it can be said that they have brought nothing but blessings. Concerning some, indeed, a case can be made out for holding that they have proved to be little else than curses ; but most inventions, it is clear, have had both good and evil results. Such is the position of the invention of mechanical transport. On the one hand, it is to the good that men can move easily and so enlarge their opportunities and experiences. On the other hand, it is mechanical transport that has made possible the rise of huge cities ; had it not been for railways, trams, buses and tubes, traffic would have seized up long ago in the centre of these cities and this would have brought their growth to an end. It is now acknowledged that on account of their mere size our larger towns are faced by difficult social, economic, and administrative problems ; evidently the origin of many of these troubles must be sought in the use made of the new transport facilities.

Dr. Liepmann is concerned with one aspect of this matter, namely the journey to and from work, and with its results, good and bad. Many people say that the time and money spent on these journeys are wasted, and it is certainly true that, other things being equal, the situation of the man who lives close to his work and can walk or bicycle to it is far happier than that of the man who must travel. Less attention is perhaps paid to the positive evils arising from daily travel. It is not merely that time is wasted ; it is not just that so many hours are lost from the day. To many, daily travel is a cause of nervous fatigue ; the zest of a new day is worn off before it has begun. More than that, no faculty can be properly exercised while travelling in crowds ; it is difficult to read, to observe, to concentrate or to contemplate with any profit. The necessity of passing so much time each day during which profitable activity is difficult or impossible encourages the habit of passing all leisure time in this way. The daily journey to work is one of the features of modern life making for depersonalisation and accounts in some degree for the inability of men now-a-days to enjoy themselves in creative activity or, indeed, in any fashion which demands more than passive response.

There is, however, no little to be said on the other side, even from the point of view of the daily traveller. Because he can travel he can choose his occupation ; his choice is no longer limited, as in the past, to whatever openings there may be close at hand. He can change his occupation if he finds that his first choice does not give scope for his talents ; he is less in the grip of an industry that does not suit him or of an employer with whom he does not get on. From the public point of view there is even more to be said on this side. In general an employer can expect that suitable work-people will find their way to his works or premises ; only in exceptional cases must an employer provide lodging. If employers had had to concern themselves with the provision of living quarters, the development of modern industry would have been hampered, and therefore the improvement in the standard of living would have gone on more slowly.

The daily journey to work is thus not an unimportant matter ; moreover, it does not stand in isolation from other social problems. Because of the texture of society, social problems, it is true, are seldom self-contained, but this problem is very closely linked with many others such as housing and the location of industry, and also to some extent with the mental and physical wellbeing of our people. As is so often the case in the field of social studies we have only a general impression concerning the facts and concerning the nature of the situation. This is not enough ; we must discover what is going on, and we must thus evaluate the results. Dr. Liepmann, who has made a start in the exploration of the situation in this country, began to work on data collected in other lands. It is to be hoped that her careful analysis and discussion of the available material will receive the attention which they deserve, and that her work will be followed by other studies as a result of which we may hope to construct a policy to guide us towards a solution of the difficult problems involved.

A. M. CARR-SAUNDERS.

24th May, 1943.

PREFACE

IN the present widespread discussion of the many aspects of physical reconstruction, the journey to work has attracted a measure of attention which was absent before the war when I undertook to write this book. The study was completed before the publication of two important Reports : that on Land Utilisation in Rural Areas by the Scott Committee, and that on Compensation and Betterment by the Uthwatt Committee. The recommendations of both Reports, concerned with decentralisation of industry and with public control of land respectively, are not without relevance to the journey to work. However, since they do not modify the conclusions arrived at in the following discussion, I have not found it necessary to include an examination of the two Reports or to make alterations in the light of their findings.

I gratefully acknowledge assistance in the preparation of this book by all who have supplied material and information. In particular, I desire to express my thanks to the Industrial Welfare Society for most valuable help in the collection of statistical data and to its member firms which answered a questionnaire, especially to those which have provided more detailed information on their employees' journeys to work. The London School of Economics kindly allowed me to use the original household investigation cards of The New Survey of London Life and Labour. The study has been approved by the University of London for the award of the degree of Ph.D. I am under an obligation to several of my friends for reading the manuscript, making valuable suggestions and helping in the preparation of the index. To Professor Carr-Saunders I am deeply indebted for criticism and advice throughout the work.

K. L.

London, June 1943.

CONTENTS

PART B: STATISTICAL INQUIRIES

CHARTS

INTRODUCTION

"IN many parts of the country masses of population move in tides of daily ebb and flow."[1] In the orbits of the large towns and in other industrial areas, travelling to work has become the normal state of things. Wage-earners and salaried employees, men, women and juveniles travel to their workplaces in the morning and travel home in the evening. The outward result of these movements is the rush-hour traffic. In addition, the journey to work has more general, if less visible, implications. It helps to expand the labour market in quantity and in quality. Mobility of labour as afforded by the daily journey widens the range of choice for the location of industry. The possibility of daily travelling materially affects the residential distribution of the population ; the journey to work is thus of great importance for town planning.

Travelling to work is a feature common to industrial countries. The terms used in different languages reflect various aspects of the complex phenomenon. The English expressions "a night- and a day population" as well as "dormitories" and "work-places" acknowledge the emergence of a "new factor in demography".[2] "Rush-hour" traffic calls to mind the hardships of travelling. The American term "commuter" refers to the individual who alternates between his residence and place of work every week-day. The German term "Pendelwanderung" dwells on the peculiarity of a movement which combines elements of migration and of a shuttle service. The French use translations of this expression, namely "migration oscillatoire" and "mouvement alternant". The Dutch word "forensen" (foreigners) points to the fiscal and administrative problems involved and hints at the split of local allegiance experienced by those who work outside their home districts.

Until recently, the journey to work with all its demographic, social and economic implications attracted comparatively little attention. But of late years, numerous practical problems arising out of the daily movements have figured in discussion of

[1] Census of England and Wales, 1921. *General Report*. p. 190.
[2] Ibid.

town and country planning and of other matters of public concern. However, this increasing interest has not been accompanied by a corresponding increase in knowledge of the issues involved. Precise information on the facts and effects of the journey to work is scarce, and there is in particular a lack of research from the economic and sociological angles. Most inquiries into the daily journey have been confined to one or two aspects of the question ; the one comprehensive evaluation, *Les Abonnements d'Ouvriers sur les Lignes de Chemin de Fer Belges et leurs Effets Sociaux*, by Ernest Mahaim, was made over thirty years ago, when some problems now prominent had scarcely arisen. It is proposed in the following study to examine the manifold aspects of the daily journey and to attempt a balanced appraisal of its position in the modern social and economic system.

The first part of the book is occupied by a general discussion of the issues raised by the journey to work. The second part provides statistical information, both on the methods of inquiry and on the results obtained. The part begins with a review of the more important investigations carried out in various countries. Subsequently statistical material collected for this study from English industrial areas is presented and analysed. A final chapter puts forward the conclusions and recommendations which emerge from the statistical examination and from the general discussion.

The study is based on peace-time conditions but takes account of such war-time features as appear likely to be of importance for future development.

PART A: General Discussion

Chapter I

DIRECTION OF THE DAILY MOVEMENTS

EVERY morning millions of workers leave their homes on their way to work. At first sight, the routes taken give the impression of a maze; but closer examination reveals certain definite groupings.

The most marked trend is from residential suburbs (" dormitories ") to business and industrial districts. This stream—and its reverse in the evening—is so dominant that, ignoring cross-currents, the whole movement has been described as " tides of daily ebb and flow ".[1] As workplaces are usually in the centre, and residential districts on the fringe, of towns, the general trend of the tide is centripetal in the morning and centrifugal at night. The outstanding example is the City of London, which " regularly expands and contracts between an insignificant night population of 13,709 and the more than thirty times as large number of 436,721 in the day ".[2]

The main trend is, however, cut across by multifarious cross-currents and counter-currents of various volume. *Cross-currents* are most conspicuous in areas with several industrial centres within daily travelling distance, such as are to be found in Lancashire, Yorkshire and the Midlands.[3] But they occur also in big towns and conurbations with a number of business and industrial districts. Thousands of people travel daily from one outer London Borough to another, as against the more usual movements between outskirts and centre; to give a random example: 1,350 inhabitants of Wandsworth travelled to work in Hampstead in 1921.[4] The Merseyside boroughs presented a

[1] Census of England and Wales, 1921. *General Report*, p. 190.
[2] Ibid., *Workplaces*, p. v. [3] Ibid., *General Report*, p. 193.
[4] Ibid., *Workplaces in London and Five Home Counties*, p. 8.

similar picture, and "between 10% and 60% of the working population of every Tyneside town went daily to work in some other town; between 9% and 50% of the workers in every Tyneside town had come in from outside".[1]

Counter-currents. In large towns which attract masses of labour from outlying suburbs, there are at the same time centrifugal movements. Inhabitants of central London districts travel to work in Reading, Slough, and Hatfield, to mention only a few instances; streams of workers leave central Birmingham each morning for Longbridge, Bournville, Fort Dunlop and other places; many Liverpool people proceed to work in Speke, but also to Manchester, Salford, and other more distant places. Manchester sends workers to Trafford Park; Glasgow to Hillington, and so forth. In all these towns routes can even be traced from the fringe of the built-up area across the city and out again to the opposite fringe; daily journeys of inhabitants of Becontree to factories on the Great West Road are a case in point. Counter-currents are not, however, restricted to large towns. In some smaller places, both daily exodus and influx are considerable and neutralise each other numerically. Chatham, e.g., had, in 1921, 3,416 inhabitants working elsewhere against 3,905 coming to work from other places. In Brentford, on the other hand, exits numbered 3,248, entries 3,074.

Yet a different light is thrown on the daily journey if viewed from the dwelling place and the workplace respectively; if viewed from the dwelling place the configuration is one of *dispersion*: the inhabitants of a neighbourhood leave each morning in various directions on journeys of very different lengths. The extent of this phenomenon, even twenty years ago, was impressively demonstrated by the specification of workplaces of the enumerated population in the Census of England and Wales, 1921.[2] A more recent example is provided by a L.C.C. census in Watling in 1937.[3] The dispersal on their daily journeys of the over 7,000 workers from this estate in the Hendon district was such that 38% worked within Hendon, a further 20% in various areas not very far from Hendon, while 42% travelled to a number of other London districts.

[1] H.A. Mess, "The Growth and Decay of Towns," *Political Quarterly*, July–Sept., 1938, p. 393.
[2] See Statistical Part, p. 113 fol. [3] See p. 22 f.

The same is happening in a unit much smaller than a residential area ; a sample inquiry in some London Boroughs [1] revealed a remarkable degree of dispersion even from individual homes where there are two or more earning members in a family.

Viewed from the other end, the configuration is one of a *conflux* at the workplace. The Census (1921) returns on " Workplaces " show this feature as complementary to the dispersal from domiciles. Actually, there is a high degree of concentration in a comparatively small number of workplaces, of employees coming from a very much larger number of scattered dwelling places.[2] The situation in this country, twenty years ago, was summed up by the Registrar-General as follows :

> Since the needs of industry and commerce have so far been best met by a concentration of the day population, while for residence and domestic purposes converse conditions are preferable, industrial areas are generally associated with a larger number of dormitory areas, and such intersection as takes place is characterised by a daily pulsation between the common industrial centre and a larger number of surrounding residential areas.[3]

On broad lines this description still holds good, although some changes have occurred since this was written. On the one hand, dormitories have been built for much more numerous populations, with the result that the numbers leaving such areas on their journey to work are considerably greater than formerly. On the other hand, workplaces have come to be more decentralised within the urban area, and there are now more very large works which employ thousands of workers. These developments have resulted in producing a more marked conflux towards the single works. Investigation of the daily journeys made by the employees of particular firms have yielded illustrative charts : hundreds or thousands of routes start at the disseminated homes of the respective firm's employees and converge at the factory gates.[4]

The terms " dispersion " and " conflux " should not be taken to imply that the daily movements of various workers from a given residential area or to a given workplace form regular patterns. Some streams of movement are short, and these are usually the most voluminous, others are long and of less volume.

[1] See Statistical Part, p. 155.
[2] See p. 116, on the relevant findings in the Industrial Region of Central Germany.
[3] Census of England and Wales, 1921. *General Report*, p. 193.
[4] See p. 200 ff.

Furthermore, the routes taken are often only in a limited number of directions. This is especially so in the case of places on the fringe of a large town or industrial area. Longbridge, for instance, which is situated in the South-West corner of Birmingham, draws its employees largely from North-East to North-West, a certain portion from the South, and hardly any from East and West. Carreras' cigarette factory, on the other hand, is located within the built-up area of London and has the lines of its employees' journeys converging from every direction. Corresponding differences exist between central and outlying residential districts.

The configuration of the routes taken on the journey to work must not be regarded as stable ; viewed over a period, the picture is kaleidoscopic. The various currents of movement vary in strength and direction ; some expand and contract periodically, others undergo less frequent but more definite changes.

Chapter II

FUNCTIONS OF THE DAILY MOVEMENTS

CLEARLY, these extensive and diverse daily movements must be prompted by various strong motivating forces. What, then, are these forces? In describing the various trends, it has already been mentioned that some journeys are due to the way in which urban settlements have grown: topographic conditions are one main cause of the journey to work. Others lie in the economic and social fields: in the structure and requirements of the modern system of society.

While the tides of daily ebb and flow between outlying dormitories and central workplaces are in general accepted as a natural consequence of the big urban agglomeration, the cross- and counter-currents are frequently considered as reflecting a lack of organisation and deprecated as a mere waste of money, time and energy. It will be seen, however, that more detailed analysis leads to a different appraisement of the position.

1 *TOPOGRAPHIC:* Co-ordination of Dormitories and Workplaces

Modern town-growth is characterised by the separation of residential quarters from industrial and business districts. The earning inhabitants of dormitory suburbs travel to work as a matter of course—the very name "dormitory" implies that it is located at a distance from potential workplaces. Factories, too, are being set up in more or less isolated localities, so as to make travelling a necessity for practically all their employees; this applies to large single works as well as to Trading Estates with numbers of smaller factories.[1] Daily travelling has enabled workers to live beyond the sometimes offensive and often unpleasant vicinity of factories, docks and offices.

[1] It is beyond the scope of this investigation to examine the whole question of the location of industry. This problem formed a major item of the task of the Barlow Commission ("Royal Commission on the Distribution of the Industrial Population"), 1940, Cmd. 6153, but our study is focused on the journey to work and discusses the location of industry mainly from the point of view of the worker.

The spread and sprawl of the towns in recent decades is a phenomenon so pronounced and so generally realised that we need not enlarge on it. Whether in the form of outlying housing estates like Becontree, which is about twelve miles from Charing Cross, and Moulsecoomb which is approximately three miles outside Brighton, or in the form of continuous peripheral expansion such as has taken place in almost every town—the housing policy between the two wars has frequently located the homes of the people miles away from their places of work. In a similar way, works have been established at considerable distances from their prospective employees' domiciles, both by individual firms (an engineering works at Hatfield), and by trading estates (Slough). The result has been that London occupies several hundred square miles of built-up area, that Birmingham covers 80 square miles,[1] and so forth. Greater London is about 30 miles across, and the distances within the built-up areas of other towns have grown correspondingly. Several factors have contributed to this development. The towns have grown in number of inhabitants and even more in number of households ; a low density standard has come to be adopted ; [2] building is carried out on " mass-erection " lines, requiring extensive building sites. The easiest way to secure large sites, however, is to get away from the town centre.

Such a housing policy would have been impossible without the spectacular growth of mechanical transport. The functioning of modern conurbations, with dormitories and workplaces divorced, is absolutely dependent upon an elaborate system of transport services. What has happened is that the technical possibility of carrying masses of people considerable distances to and from work has been utilised, and more or less lengthy journeys have become the daily routine of millions of workers, without much consideration being given to the economic and social problems involved. The question as to where the prospective inhabitants of the new housing estates would have to work, and how they could travel, has not usually been taken into account ; the efficiency of modern transport has been relied upon to provide any necessary facilities for conveying people to and fro. Distance did not seem to count—modern transport by rail and

[1] H. J. B. Manzoni in *Journal of the Town Planning Institute*, Jan.–Feb., 1940, p. 41.
[2] See p. 92.

by road would cope with it. This indiscriminate use, or almost deliberate misuse of transport facilities was described by F. J. Osborn, when he complained of " the transport industry which began by making it possible for city people to ' live out ' and ' work in ' and seems to end by making it possible to work anywhere and live nowhere ".[1]

In addition to these centripetal journeys to work, there are centrifugal ones, from homes in the old central districts to new factories on the outskirts of the towns. It is too early yet to see to what extent this tendency of " living in " and " working out " will develop. It is conceivable, however, that it may remodel town life considerably: after all, men are remaining in, and flocking to, the town largely in order to enjoy its facilities for a fuller life, in such spheres as social intercourse, education, amusement and shopping. At the same time, an arrangement which locates the residential quarters in the centre of the town and industry at the fringe provides workers and employers alike with good opportunities for choosing suitable jobs and labour respectively. Easy access to the open country would be a necessary feature of such a town plan.[2]

It has been due to the attitude of neglecting the implications of the daily journey that " there has been no attempt towards securing development with proper correlation of workplace and home ".[3] " There is little consultation between the transport agencies and local authorities in the matter of disposition of housing estates." [4] In Greater London the main new housing estate was developed in the East of the town, while most of the new factories were built in the West. The result has been daily journeys across the whole width of London, from Becontree to the Great West Road, for instance. A similar situation arose in Glasgow, as the result of non-co-operation between various departments, " the Special Commissioners having located Hillington, their trading estate, five miles on one side of Glasgow, and the local authority in Glasgow building their housing development five miles on the other side of the City ".[5]

[1] *Transport, Town Development and Territorial Planning of Industry*, New Fabian Research Bureau, 1934, p. 4.

[2] See p. 108.

[3] Sir G. Gibbon, Barlow Commission, *Minutes of Evidence*—hereafter referred to as " Barlow Commission, *Evidence* ", p. 863.

[4] Major H. E. Crawford, ibid., p. 688.

[5] Sir Montague Barlow, ibid., p. 874.

2 *ECONOMIC AND SOCIAL*

The social and economic importance of the daily journey consists in its contribution to the mobility of labour, " the very breath of life to modern industry ".[1] The primary manifestation of mobility of labour is, of course, migration : workers move from their old homes to the neighbourhood of new workplaces. There is a continuous migration of individuals and of families towards new centres of economic activity, thus adapting the geographical distribution of the population to the changing needs of industrial development. This re-orientation is taking place on broad lines and over long periods, but it entails incisive ruptures of human and of economic life.[2] The smooth working of the social-economic organism requires adjustments of a less radical kind ; the flexibility of the industrial structure is assisted by the daily journey to work.

Apart from facilitating economic change, the journey to work fulfils important functions also in a stationary society : daily travelling widens the labour market, making easy the choice of employment to the employee and the choice of employees to the employer ; this increases the independence, in both private and working life, of the wage-earner ; firms too are rendered less dependent on local labour in recruiting their workpeople. Further, daily travelling helps to preserve the family unit, by making it possible for various earning members to work in different localities while maintaining home life in the domestic circle. The daily journey thus serves to co-ordinate factories and offices with their varying labour requirements on the one hand, and the human beings offering their labour and grouped in families, on the other hand.

a Essential Condition of Large-Scale Undertakings

The emergence of large-scale manufacture as the most economic form of production in various branches of industry has led to the development of huge plants, employing many thousands of workers, mainly in the iron and steel, engineering and chemical industries and also in the retail trade in the form of department stores. It would be physically impossible, e.g. to house the

[1] *Balfour Committee on Industry and Trade*, Final Report, 1929, Cmd. 3282 p. 235.

[2] See A. D. K. Owen, " The Social Consequences of Industrial Transference " in *The Sociological Review*, October, 1937.

20,000 employees of an engineering works in the vicinity of the factory: the great majority of such workers being men, they and their families constitute by themselves the population of a medium-sized town. The community would be further increased by the auxiliary services which are necessary for such a number of inhabitants; the area covered by houses would render illusory the nearness between home and workplace. Moreover, there would be the social and economic disadvantages of a one-industry town. Even for a town of the size of Oxford, apprehension is felt about the one-sided type of employment offered and about serious results of a depression in the motor-car industry, upon which alone its industrial population depends. Daily travelling by the workers has thus become necessary to secure the concentration of labour in plants of the size demanded by technical and economic considerations.

b Contribution to the Mobility of Labour

i WIDENING OF THE LABOUR MARKET IN GENERAL

From the point of view of the worker, the principal significance of the journey to work is that it extends the market in which he is able to offer his labour and thus enlarges his economic independence. He is not obliged to accept a job from those firms which he can reach by walking from his home. Moreover, daily travelling puts at his disposal a more varied choice of employment in a broader range of occupations and industries, to suit his inclination, training and skill. He can more easily change his employer if he wants to, without having to move his home to another locality, with all the cost and inconvenience this would involve; this is specially important for those who own the houses in which they live.[1] The possibility of daily travelling to work reduces the danger of unemployment—more will have to be said on this point.

In the private sphere, too, the worker derives a greater amount of independence from daily travelling; the fact that he can choose his place of living, to a large extent uninfluenced by his place of work, means an increase of personal freedom. This is most clearly demonstrated by the conditions as they exist in certain American "Company Towns", where not only the housing accommodation but also utility services and amenities

[1] See p. 17.

are provided by the one company on which the inhabitants depend for employment. In such towns, there is hardly any side of the workers' life which is not, in one way or another, under the influence of the dominating employer. In this country, the system of paternalism under which the employer is at the same time the landlord belongs now, on the whole, to the past. Even to-day, however, it is not without significance that the worker can choose his place of residence beyond the sphere of an employer who is influential socially and politically.

A development should be mentioned here which operates to prevent the worker from becoming more independent in the choice of his dwelling place. At the same time when the tie binding the employee to the vicinity of his work is loosened by transport facilities, his freedom to choose where to live is endangered through certain consequences of the housing policy in this country. Since the last war, local authorities have become almost the sole agents for building houses for the poorer sections of the community.[1] This fact influences the location of individual workers' homes in two ways : first, for various reasons, mainly administrative and financial, municipal building has in some towns been concentrated in a few large housing estates. Secondly, the monopoly of subsidised building in the hands of the local authorities has implied undesirable restrictions in the allocation of new dwellings : as a rule, a man is dependent for being housed or rehoused on the local authority in whose district he has been living. This creates difficulties in moving from one London borough to another as well as in inter-urban migration. The following is a case in point : when the Ford Motor Co. moved their works from Trafford Park to Dagenham, the workers coming with the firm could not be accommodated in the L.C.C. estate at Dagenham, since these houses were reserved for families who had been living in the London County area for at least two years. It is obvious that these features of municipal housing policy restrict the workers' freedom to select the place of their domicile according to their personal circumstances, and thus

[1] This organisation of subsidised building is peculiar to British housing policy. With the exception of the fully socialised provision of houses in Vienna and, of course, Russia, the continental countries have developed bodies intermediate between the public authority, on the one hand, and the prospective tenants, on the other hand—see the author's article, "Public Utility House Building in European Countries", in the *International Labour Review*, 1929, Vol. XX, 1 and 2.

reduce the mobility of labour which is recognised as an essential factor of modern industry and of the worker's independence.

The benefits which the employer derives from the extension of the labour market brought about by the workers' habit of daily travelling correspond to those accruing to the worker. The possibility of selecting his hands from an area wider than the immediate vicinity of the factory premises makes it easier for a firm to find "the right man for the right job". This is very apparent in the case of skilled and otherwise specialised jobs. Standard Telephones, e.g., draw the male employees of their New Southgate factory from all over London, as it would not be possible to find the necessary number specialised in light engineering within walking distance of the works.[1] Another example is the employment by the motor industry in Birmingham of people from the Black Country for certain hard work to which they are traditionally accustomed and which they are better able to do than townsfolk. The workers concerned, on the other hand, obtain higher wages in Birmingham than those prevailing in their own district. For the recruitment of several hundreds of stokers, the Leuna Works (Central Germany) went as far as to the Mansfeld District, over 30 miles away, where the decline of the copper mining industry had set free labour suitable for this hard work. Special workmen's trains were run for these men. Incidentally, an interesting feature was that, since the Reichsbahn-Gesellschaft asked for a minimum of 500 passengers for these trains, a number which surpassed Leuna's demand for the type of worker to be drawn from the Mansfeld district, the firm recruited other workers from places along the route, in order to make up 500. The rather long journeys of these complementary men had thus an origin connected neither with the firm's nor their own actual requirements—a situation whose causes can seldom be so clearly traced but which may be assumed to prevail in many cases. The history of the Leuna works provides also an illustration from a different aspect of the importance for the employer to be not dependent on local labour. In 1920, the works was for several days occupied by rioting workers—thereafter, the firm's housing scheme was altered and only a restricted number of watchmen, foremen, and key-workers accommodated in the immediate neighbourhood of the works.

[1] See the Statistical Part, p. 145.

This occurrence may explain why also in less extreme cases a firm may favour daily travelling by its employees, out of dislike for having an isolated factory surrounded by the homes of masses of dismissed or otherwise dissatisfied workers. "In the opinion of the directors there are many disadvantages in a Joint Stock Company being landlords as well as employers" was the answer by an important factory in this country to the question about firms' housing schemes for their employees. From both the employer's and the worker's points of view, firms' housing schemes are to be considered as possible causes of friction in industrial relations.

ii FACILITATING ECONOMIC CHANGES IN PARTICULAR

The daily journey's contribution to mobility of labour is especially important in the process of economic change such as occurs in expansion and decline of industry, short-distance migration of factories, seasonal fluctuations and labour-turnover in general.

Expansion of Industry.—The industrialist who plans to erect a new works finds labour among the principal location factors, together with raw materials, site and services and markets. Generally, location will be based on a combination of these various factors. The factor "labour" can be taken into account in three ways; either the new works has to be situated in the vicinity of a pool of suitable labour, or new housing accommodation adjacent to the factory has to be provided for the prospective employees, or the problem may be solved by relying on the workers' daily travelling. The first method must often conflict with other main location factors (sites, e.g., are most expensive in densely populated neighbourhoods, and the transport of raw materials to such premises may be expensive). The second method (housing the required workers near the new factory), if undertaken by the industrialist himself, involves initial capital outlay to an extent which may be prohibitive. To leave, on the other hand, the construction of houses to private enterprise or to local authorities means a postponement of the start of production, a delay which it might be essential to avoid.[1] Daily travelling

[1] "The success of many new enterprises depends on the speed with which it is possible to start production." P E P *Report on the Location of Industry in Great Britain*, 1939, p. 90.

by the workers is the expedient to solve this dilemma. Though the firms, as will be shown below, have in many cases to subsidise their employees' transport during the initial months or even longer, those expenses are incomparably lower than would be the costs of erecting dwelling houses for them.—To some extent, this function of the daily journey may lose in importance. Light mobile industries employ a growing proportion of earners, and trading estates providing suitable premises for small new factories may come to be located within easy reach of residential areas. But this is only a partial solution.

It is in fact a common experience that new and expanding works have high proportions of their employees coming from considerable distances, as against more local recruitment by old-established firms. This is especially marked where a huge plant is opened in a rural district, the location being determined by factors other than labour. The Leuna Nitrogen Works, for instance, the construction of which was begun in 1916, are dependent on the vicinity of bituminous coal-pits, and strategical reasons accounted for the location in Central Germany. Only secondary consideration, consequently, was given to such factors as labour supply. Without the device of daily travelling on the part of nearly 20,000 employees, production at the Leuna Works would have been held up for years; in the early 1920's, the erection of the necessary number of houses would not only have meant an enormous capital outlay but, under the post-war conditions in the building industry, have been uncertain of achievement. Similar conditions prevailed at Eindhoven, Holland, during the rapid growth of the N.V.Philips Gloelampenfabrieken in the 1920's; the town of about 80,000 inhabitants could not supply the 20,000 employees wanted by the firm; consequently large numbers, including many women, were drawn from other places far and near. The same problem was faced by the De Havilland Aircraft Company on opening their new works, and they too were helped by being able to draw their employees from far afield. When a big engineering firm undertook to build a "shadow" aircraft factory under the Government's rearmament programme in 1936, the site chosen was on the firm's own premises. The weight of the location factors in favour of this place (readiness of site and services, common management with the existing huge works) must obviously have outweighed other factors, including labour supply, which would have recommended a different location ; it proved very difficult to carry the required number of additional workers to the factory, but ultimately daily travelling was made possible.

Short-distance migration of Industry.—The same factors operate in the removal of a factory to another district. During the 1920's and 1930's quite a number of firms have seen fit to move their works from the centre to the fringe of large towns, or even beyond, where production and working conditions seemed superior. As a rule, the chosen site was beyond walking distance from the homes of the employees and mostly without nearby adequate housing accommodation. Even if there should be a numerically sufficient supply of labour in the new neighbourhood, the firm could not consider relying altogether on this labour, thus parting with its old workers, whose experience, training and loyalty are an asset, imponderable but none the less important, to the prosperity of the firm. On the other hand, insistence on its old personnel being housed in the new district would place the migrating factory in the same situation as an entrepreneur starting production afresh. The daily journey solves these difficulties. There is ample statistical evidence that in works transferred to their present site as long as five and more years ago, the residential distribution of the employees still largely reflects the former location of the factory and entails daily journeys for the workers concerned.[1]

Reorganisation of multiple Plants.—A development similar to the transplantation of entire factories is the concentration in a single place of the various local branches of one firm ; the daily journey is important also in this connection, as shown in the case of the London Passenger Transport Board. In the course of its formation out of numerous local transport agencies, the Board took over repair shops and garages scattered over a wide area. In the subsequent unification, " inadequate and inconveniently situated garages acquired by the Board have been closed . . . Chiswick Works are being radically replanned and enlarged to enable (all) the buses and coaches to be efficiently overhauled . . . The work of maintaining and overhauling trams has been centralised at Charlton Works, while trolley-buses are dealt with at Charlton Works and Fulwell Depot. Eight repair-shops have,

[1] See Part B, pp. 134, 141, and 145.

in consequence, been closed."[1] This meant a divorce of home and workplace for the mechanics and engineers who lived near the closed garages and repair shops where they had been employed. If they had not been able to travel to the new and distant workplaces, either the Board would have lost these experienced men and the men their jobs, or the latter would have had to move nearer to the new workplaces and thus forsake their residences.

Continuity of residence is generally desirable in the interest of private and of community life. For those who have bought their houses it is furthermore an important financial consideration. Genuine ownership as well as the hire-purchase system bind a man to his residence, for often the house cannot be sold without loss. This tie either encroaches upon the mobility of the owner-occupier, if it prevents him from accepting an advantageous but distant job, or else, if he does accept it, leads to daily journeys of excessive length. For this reason alone, if for no other, the purchase of a house, so emphatically encouraged from various quarters, cannot be regarded as beneficial to the wage-earner. The need is for plenty of well distributed dwellings to let, to help families which want to move nearer to new workplaces.

In addition to the more far-reaching and fundamental changes there are less drastic but more frequent changes in which the journey to work plays an equally important part.

Seasonal Fluctuations.—In such regular occurrences as seasonal trade fluctuations daily travelling acts as a shock-absorber. In the motor-car industry, for example, there are slack months in the year when many hands are temporarily dismissed ; they are re-engaged with the start of the annual upswing of production. In the meantime these workers take up other temporary jobs. Many of the large motor works are at some distance from other factories, and only by daily travelling between home and workplace can unemployment be avoided and the firms enabled to recruit the full number of men needed during the peak period of their activity.

The seasonal character of employment is very marked in *agriculture*. In fact, on the Continent many workers alternate between agricultural and industrial employment, in the course of the year. In the rural districts of Wurttemberg (South-West

[1] L.P.T.B., *5th Annual Report*, pp. 11–12.

Germany) the law of inheritance led to a splitting up of farms to such a degree as to make the individual plot too small to support a family. The result often was that farmers took to industrial work during the winter months without giving up domicile on the farm; comparative nearness of factories and facilities for travelling thus enabled them to supplement the insufficient yield of their shrunken farms.[1] In Belgium, too, Mahaim's investigation showed a large proportion of daily travellers to be farmers taking up work in factories or mines as a seasonal job; farming was not discontinued but had ceased to yield enough for the livelihood of the family.[2]

Labour-turnover.—Apart from workers who are temporarily dismissed during the slack season, there are unstable elements who do not remain long in any job. Statistical information on labour-turnover is scanty; the ratio varies from industry to industry, between areas and from one firm to another. That the figures are sometimes quite considerable is shown by data given for the Morris Cowley Works; the aggregate of workers entering and leaving the firm during 1935 accounted for 75% of the total number employed at the beginning of the year,[3] while the actual increase of the personnel was only 11·7%. Whether the job is of a non-permanent nature, or whether frequent loss of employment is due to personal reasons, daily travelling makes it easier for those dismissed to find new work.

Decline of Industry.—During a period of depression of an industry or when a single works is shut down, the daily journey is of outstanding importance for the workers affected. Where other employment is within daily reach the people concerned are spared the alternative of long unemployment or uprooting of the home. Usually, the dismissed workers look for new jobs individually and find them by infiltration, as it were, into other works. Longbridge, for instance, has numbers of such men from the Black Country on the pay-roll.[4] We know, however, of cases in which a common solution has been found by arrangement for daily travelling, and thus the social and economic consequences of protracted and extensive unemployment have been avoided. In the Reports on the Depressed Areas[5] it is stated

[1] *Württ. Jahrbücher für Statistik und Landeskunde*, 1902.
[2] See p. 123.
[3] Including seasonal fluctuation.
[4] See p. 147.
[5] Cmd. 4728, 1934, p. 200.

that " one example, specially worthy of note, is travelling of miners resident in hamlets adjoining closed pits to other pits, often belonging to the same coal-owners, where housing is not available ". While the Report mentions this as an outstanding instance, it is the usual policy in Belgium : after the exhaustion of a coal-mine the displaced miners are daily conveyed to work in another pit ; long-distance trains at cheap rates are run for this purpose.

The problem becomes less manageable when unemployment is widespread as a result of the decline of the staple industries of a whole area. The non-existence of alternative labour markets within daily travelling distance has been a contributory cause of the dereliction of the " Special Areas ". In Scotland, e.g., " the basic industries . . . are particularly interdependent, and the sharing of prosperity or depression by the whole area is inevitable on this account ".[1] All the more is it important that different industries within daily reach should give opportunities for other employment to the workers in areas stricken by the decline of their staple industries.

The usefulness of the journey to work is not restricted to a system of private enterprise governed by competition. A planned society would equally be faced with the conflicting needs of economic change and of continuity of community life. In any dynamic society " the prompt and continuous diversion of labour and capital into relatively new types of production is an essential condition for maintaining a satisfactory rate of material progress as well as for avoiding chronic relapse into depression ".[2]

iii PRESERVING THE FAMILY UNIT—THE CASE OF THE SECONDARY EARNERS

A considerable number of daily journeys are due to the impact on the family unit of the organisation of modern industry. Homes and workplaces have come to be not only separated in space but also different in structure. The labour supply offered by a family may consist of husband and wife, or, more often, of father and one or more children, sometimes supplemented by other relatives. In the modern labour market, however, this

[1] Ibid.
[2] Allan G. B. Fisher, *The Clash between Progress and Security*, 1935, p. 204.

family group is in demand not as a unit but as individual workers.[1] The composition of labour in workshop and office is determined by types and grades of occupation, and little thought is given to drawing people from the same home. Reflecting the immense variety and differentiation of modern industry, the demand for labour in most firms is to a high degree specialised. Requirements vary as regards physical strength, skill, training and adaptability, and consequently as regards sex and age. Many firms employ great numbers of one special type of labour. In these circumstances it is obvious that the various members of a family often fail to obtain work in the same neighbourhood. The daily journey, however, brings jobs within reach of those members who are unable to find suitable work near home and thus helps considerably to preserve the family unit. The process of disintegration of the family has been going on for many decades ; it would have proceeded much further without the facility of travelling to work. Assuming, in the first instance, that the family's home is near the householder's job, it is the secondary earners who have to travel to work. The average number of earners per family was found to be 1·72 in the Survey Area of the *New Survey of London Life and Labour,* in 1929.[2] This means that for every ten principal earners (mainly male householders) there were over seven earners other than the householder (mostly sons and daughters) ; in other words, 41·8% of all earners living with their families were not in a position to move their home to the vicinity of their work.[3]

[1] In non-factory work, certain cases of the family as a working unit survive to this day : some peasants, craftsmen and shopkeepers continue to employ all working members of the family in their business on the premises of the home. On the Continent, the habit is still alive of a whole family being hired and accommodated as a combined labour force for agricultural work. In this country, married couples to " live in " are a combination in demand as gardener (or chauffeur) and cook (or housemaid).

A scheme to create identity of living and working units, such as had existed in the family household of former ages, under the conditions of factory production, is to be found in Fourier's *Phalanstère* : a phalanstère was conceived as a community for living and working of 400 families in a local unit. In all such cases, real and schematical, a daily journey to work is eliminated in principle.

[2] Vol. VI, p. 34 : Average working class family, omitting families without earners ; including as " families " groups of one or two earners. The average number of earners was 1·75 per household in the inquiry by the Ministry of Labour into " Weekly Expenditure of Working-Class Households in the United Kingdom in 1937–8 ". *Ministry of Labour Gazette,* Dec. 1940, p. 304.

[3] It would be interesting in this context to examine the conditions as to the journey to work of those persons who leave their home in order to live nearer their workplace. They form an unknown proportion of the *lodgers and boarders* i.e. the " many

Girls in particular often lack the opportunity for work near their father's workplace. This is most conspicuous where there is a daily exodus of female workers from an area which attracts male labour from outside. The main example is that of mining districts, which offer occupations for men only, while women and girls travel to work, for instance, in the textile and clothing industries. Such conditions prevail in the West Riding, where Leeds is the centre of attraction for female labour from the surrounding mining villages. In the case of Bootle, a net daily influx of men is contrasted with a net daily outflow of girls.

> While the large modern docks, engineering and ship-repairing works, timber yards, tanneries etc. . . . provide occupation for a very large body of men, drawn in part from outside, the large shipping and commercial offices which are concentrated in Liverpool, and the large stores and shops in that city, attract a considerable number of female workers from Bootle.[1]

In Litherland, Lancs, females accounted for over half of

unattached men and women whose circumstances necessitate their living away from their own homes and who have no intention or opportunity of setting up as housekeepers themselves ", (Census of England and Wales 1931, *Housing*, p. xl). But the information about them is scanty and incongruous. "Exactly what numbers or proportions of the 839,000 shared dwellings in England and Wales are occupied in this way cannot be stated with precision . . . though . . . it may reasonably be inferred that the proportion is a large one." A more detailed analysis of the Census 1931 results was made for certain controlled Post-War Housing Estates in London (Becontree), Birmingham, Liverpool and Manchester (Ibid., p. xlix foll.). This resulted in proportions varying from a rate of about one lodger in every nine families in Becontree to one in every six families in the Manchester sample.

Unfortunately, no comparable figures are available for the older parts of the towns. The *New Survey of London Life and Labour* (Vol. III, p. 37) and the *Social Survey of Merseyside* (Vol. I, p. 193 ff.) give information on lodgers, but the findings —about one lodger in every 40 working-class families in Liverpool and about one in every 80 such families in the Eastern Survey Area of London—are based on different definitions. In London, only those who board with the family are counted as lodgers, in Liverpool also others. The Census records such relations as brothers and brothers-in-law as lodgers ; the two Surveys do not. The main difference is in regard to sub-tenants ; they are, as separate lodger families, included in the Census specimens of new housing estates, but excluded from the more comprehensive samples of the two Surveys ; sub-letting is, however, much more common in the inner than in the outer districts of a town (see *New London Survey*, Vol. III, p. 52).

An investigation of the local distribution of sub-tenants and lodgers would help to show in how far consideration of the journey to work influences the choice of where to live. Do lodgers (in a wide sense) live nearer their work than workers in general ? From where do the lodgers come, from other districts of the town or from more distant places ? In the Liverpool sample, more than two-thirds were natives of Merseyside. How is the number and composition of lodgers affected by the opportunities for employment for various types of earners ? A " factor which goes some way to explain the very small number of women lodgers (in Liverpool) is the lack of industries likely to attract women to Merseyside from other parts of the country ".

[1] W. Hewitt, *Workplaces and Movement of Workers in the Merseyside Area*, p. 19.

the daily influx, as against less than one-quarter of the daily exodus.[1]

Juveniles, too, join in the daily movement, but no statistical documentation is available regarding their participation. That the extent is not inconsiderable is reflected in the following statement by the London Regional Advisory Council for Juvenile Employment.[2]

> For many years local committees for juvenile employment have been deeply concerned at the comparatively high cost of travelling in relation to wages earned by juveniles resident in the suburbs and employed in Central London . . . The greatest demand for juveniles lies in the central and western half of the London Region, while the largest supply of juveniles is in the eastern half . . .

A sample inquiry in four London boroughs[3] shows about the same proportion of boys and girls using means of transport on their journey to work as in the case of adults.

This inquiry indicates further that the daily journey facilitates the social rise of a family ; by being able to travel to firms which provide proper training, a docker's or roundsman's children have access to skilled occupations often not to be found in the vicinity of their father's workplace. Many household cards of unskilled workers had entries for children, including grown-up sons and daughters, who travelled to other parts of the town where they were apprentices or had posts of higher standing, and sometimes higher income, than their fathers who worked within walking distance from home.

In other cases the situation is reversed, that is to say, the householder travels to work, while his wife or children find local employment. This happens frequently in outlying suburbs ; many factories have in recent years been located on the fringe of large towns, largely factories which employ predominantly unskilled labour ; they offer employment to women and juveniles living on nearby housing estates, but have no demand for skilled men. Such conditions are clearly visible at Watling, the L.C.C. housing estate in Hendon ; of the householders, little over one quarter worked in Hendon, as against over one half of the secondary earners ; on the other hand, more than one half of

[1] Census 1921. *Workplaces*, p. 9. [2] *First Annual Report* (1935), p. 3.

[3] Based on the household cards of the *New Survey of London Life and Labour*, see p. 155 f. below.

the householders travelled to distant boroughs, as against not much more than a quarter of the secondary earners (the balance being made up by persons working in intermediate districts surrounding Hendon).[1] However, the advantage of living near their work is for juveniles offset by the fact that their employment is of a blind-alley type ; the demand for their labour ceases when they come into their late 'teens. "For Watling youngsters, Hendon is a place where one can earn money, provided one is young enough, and London is a place which harbours prospects. . . . Those adolescents who can afford ambitions, and who want a career rather than a job, do not work near by ; they travel to London each day."[2] The journey to work thus helps to prevent social decline by remedying such dangers as are found, for instance, in Cumberland, with its lack of a supplementary labour market within daily travelling distance, namely "the almost complete absence of openings for juveniles, which is crowding children into blind-alley occupations, endangering their whole economic future".[3]

[1] Ruth Durant, *Watling*, p. 13.

PLACE OF WORK OF ALL PEOPLE AT WATLING IN 1937
(L.C.C. Census of the Estate)

Districts (of workplace)	Tenants Number	Tenants %	Others Number	Others %
Westminster and Chelsea	225	5·6	69	2·3
Kentish Town, St. Pancras, Islington	687	17·2	227	7·5
City, Holborn	552	13·8	203	6·4
Paddington	554	13·7	344	11·1
South London	80	2·0	15	0·4
	2,098	52·3	858	27·7
Hendon	1,055	26·3	1,681	54·3
Willesden and Marylebone	501	12·5	396	12·7
Hampstead and Highgate	170	4·2	108	3·2
Finchley and Southgate	70	1·7	20	0·6
Acton and Southall	72	1·7	54	1·5
Other Districts	66	1·3	—	—
Total	4,032	100	3,117	100

[2] Ibid. [3] *Reports on the Depressed Areas*, 1934, p. 41.

Apart from enabling the worker to rise in the social scale, daily travelling also widens the choice amongst callings of equal social standing. In the modern economic system in which labour has become so highly specialised, it is important that the entrance of a young person into working life should be suited to his individual capacity. Vocational guidance endeavours to steer boys and girls into careers suited to their special aptitudes. This refinement in the use of man-power can be turned to use only if openings of various kinds are available to the inhabitants of all districts. This opportunity is often lacking in a small town or poor borough, but daily travelling to a more distant workplace overcomes this difficulty and makes the choice of the right opening possible, whilst the place of residence is not robbed of its promising young citizens.

At the same time, the family as a whole benefits from the varied employment of its members. It is economically safer for the family not to have all its eggs in one basket, i.e. not to depend on one industry which may decline while others prosper. Domestic life, moreover, is enriched by a variety of occupational interests among the family.

If a severance occurs between residence and workplace, either by a transference of the factory or by removal of the family home, young persons are frequently willing and able to change their jobs so as to live again near their work. Older employees, however, stick to their jobs ; it has become increasingly difficult for men over 40 and for women even over 30 years of age to find new employment. Many older employees, moreover, have gained special experience in their jobs and enjoy confidential and responsible positions, and in some firms also can look forward to a pension if they stay on. This is made possible for them by travelling to work.

In conclusion it can be stated that everyday working life is full of instances in which the journey to work is serviceable to the worker ; daily travelling helps him in getting the right job, in retaining a suitable job, in combining two intermittent jobs, and in changing over from declining to flourishing industries. But for the daily journey, breaks in the worker's life would be much more frequent through repeated migration of individuals and of families.

In the profound fluctuations of the modern economic system, the daily journey is an important means of giving the necessary flexibility to the industrial structure and thus mitigating the impact of change. Whenever this flexibility has been lost, there have appeared areas of stagnation and decay such as the Depressed Areas. The daily journey is certainly no panacea; but the isolation of an area, without access to alternative labour markets for the principal earner as well as for other members of the family, is without doubt one factor making for economic depression.

By increasing the mobility of labour, the daily journey is not only in the best interests of employers and workers, but it also fosters the progress and prosperity of the economic life of the nation as a whole. Daily travelling to work enables the country to make the best use of its man-power; it helps to direct workers to those points where they can most effectively be put into action.

Chapter III

THE PRICE OF THE DAILY JOURNEY

I *THE COSTS IN MONEY*

a The Burden on the Community as a Whole

i An item of the costs of production in general

FOR reasons which have been discussed in the preceding pages, travelling to work is in principle an essential feature of the present social and economic system. It is, therefore, an error to regard the cost of this transport merely as a wasteful addition to the necessary cost. It is characteristic of the modern economic structure that the cost of distribution forms a considerable part of the cost of production in a wider sense ; and this applies to the transport of labour as well as to that of raw materials and of finished goods. Nor is it a valid argument against the economic soundness of transport to and from work that vehicles required for the peak traffic lie idle during the greater part of the day ; similar periods of idleness occur in other services which are governed by the rhythm of nature or of human life ; agricultural machinery is an example of the first, restaurants of the second type of periodicity.

The capital invested in transport, both public and private, for the purpose of the journey to work has to be conceived as a constituent part of the country's industrial equipment, and the running costs of railways, buses and trams as an item in the cost of production. Beyond this general fact, however, little is known of the importance of the cost of daily travelling as an item in the national economy. Under the present complicated economic organisation, the transportation of labour to the workplace naturally costs more than formerly—but what is actually the nation's bill of costs for the journey to work, and what proportion of the total cost of production is to be regarded as normal or desirable ? Neither question can at present be answered satisfactorily. More research is needed on these points. Until further information is available, recourse must be had to some illustrative data, in order to obtain an approximate idea of the magnitude of the costs involved.

First, with regard to the capital sunk in transport equipment: the railways were estimated in 1928 to account for 8·6% of the national capital.[1] This figure, however, includes all passenger traffic and also goods traffic. On the other hand, there is no comprehensive estimate of the capital invested in road traffic.[2]

Writers on the economics of railway operation agree that it is scarcely possible to separate the capital and working costs applicable to goods traffic and passenger traffic respectively. It is also difficult to single out the journeys to work from the total passenger traffic. For while practically all journeys on season and workmen's tickets are travel between home and workplace, many other "bread-and-butter journeys" are paid for at standard rates and are, therefore, not distinguishable in the statistical returns. On the London buses, e.g., no season or workmen's tickets are issued. Buses, however, account for over 50% of the London passenger traffic (in numbers of journeys).

The importance of the daily journey varies between different railway lines, but on many of them it has become the backbone of passenger traffic, and, on the whole, the journey to work accounts for a considerable proportion of the total journeys. The adaptation of the transport services for the daily conveyance of earners between home and workplace is, of course, specially marked in the case of urban and suburban traffic, as is shown by the following data for London and Birmingham.

On ordinary weekdays, three-fifths of the journeys on the system of the L.P.T.B. are made during the six hours before the start and after the cessation of work respectively.[3] While this figure includes certain numbers of travellers for other purposes, such as children on their way to school and shoppers who return

[1] Sir Josiah Stamp, *National Capital*, 1937, p. 234.

[2] A certain indication of the relative importance of passenger transport as an industrial undertaking is given by the fact that the London Passenger Transport Board, which caters for over 85% of the public local passenger transport in the London Transport area, employed, in 1939, a staff of 86,456. (*6th Annual Report*, p. 21.) This is 2·3% of the 3,767,000 persons occupied in Greater London in 1931 (Census Report), and a somewhat smaller proportion of the larger number employed in the total London Transport area.

[3] L.P.T.B., *5th Annual Report*, p. 30.

home after the closing of the shops, part of the journeys to work, notably by second-shift workers, are made at other times of the day. It is estimated that roughly two-thirds of the Board's total traffic is business traffic,[1] consisting mostly of the conveyance between homes and workplaces. Similar conditions prevail in Birmingham, where " no less than 69 % of the rolling stock lies idle between the hours of 9.30 a.m. and 4.30 p.m ".[2]

Next to the question of the aggregate cost of daily travelling, the consideration arises as to the distribution of the burden, a subject on which more is known. This is not the place to discuss the part played by local authorities in the provision of transport facilities, principally their responsibility for roads and streets, and municipal tram and bus undertakings. One measure, however, requires fuller discussion, as being specifically designed to facilitate the journey to work, namely the concern taken by the State in workmen's trains.

ii PUBLIC CONCERN IN WORKMEN'S TRAINS

On the whole, it is left to the two parties in industry, and mainly to the workers, to pay the cost of the workers' transport to and from work. Their respective shares in the expenses will be discussed in the subsequent sections. To some extent, however, these costs have been made a concern of the community : in this as in other countries the State has intervened and has assumed a certain responsibility for the cost of the journey to work. Cheap travelling has been found to be essential for industrial development. Consequently, as normal fares proved to be too high for the lower-paid worker, specially reduced workmen's tickets have become a common feature in industrialised countries. The term " normal fares " must in this context be understood to include season tickets issued for non-manual workers ; these, although reduced as compared with ordinary fares, are yet remunerative to the railway companies. Workmen's tickets, on the contrary, are commercially not profitable, and their issue is, therefore, dependent on State intervention.

The manner in which this measure of economic policy has been put into effect has differed in different countries. The

[1] Mr. Frank Pick, *Evidence* before the Barlow Commission, p. 367.

[2] A. C. Baker, " Problems for Municipal Transport ", in *Modern Transport*, June 1932, p. 13.

outstanding example of Continental conditions is Belgium, where the main railways are State-owned. When, in the early days of industrialisation in Belgium (the 1860's), the factories and mines needed cheap labour in excess of local supply, the parliament decided upon reduced fares for manual workers, and by administrative decree the State railway was directed to issue *abonnements d'ouvriers* at very low rates.[1] The railway being a public undertaking, operating losses that might arise from the rebates on workmen's tickets were borne by the taxpayer. There has been controversy as to the origin of the measure : E. Mahaim [2] maintained that the Government took the initiative, while E. Vandervelde [3] stressed the pressure brought about by the employers, who had a particular interest in low travelling costs for their prospective workers, at a stage of industrial development when the factors determining the wage level were not yet fixed. In the first decade of this century, 42 % of all passenger journeys on the Belgian State Railways were made on workmen's tickets ; after the last war, even more.

In Great Britain, where the railways are owned by private enterprise, it has been more difficult to procure cheap transport for workers. During the first decades of their transport monopoly in the nineteenth century, the Railway Companies discouraged even third-class travel, not to speak of travelling at sub-standard fares. Workmen's trains were obtained from the Railways in return for concessions made by Parliament.

The origin of the workmen's trains in this country (and in fact in any country) dates back to 1861. They were introduced as a more or less incidental measure to meet a special situation in central London, not as a general policy of cheap transport. The case is of particular interest to-day, for the reason that it contains an element of town planning. The construction of railway termini in built-up urban areas entailed the demolition of houses, often of working-class houses. It was the practice of Parliament, in granting powers to railway companies, to impose on them the obligation to rehouse the evicted people within one mile of their previous homes. When, however, Parliament was asked for powers to build Liverpool Street Station it would

[1] See p. 122.
[2] *Les Abonnements d'Ouvriers sur les Lignes de Chemins de Fer Belges.* 1910. p. 10.
[3] *L'Exode Rural,* 1903 p. 130 seq.

have been a most onerous obligation to rehouse the ejected families in the congested neighbourhood. Instead, the Railway Companies were put under the obligation to provide cheap transport for the displaced workers between certain residential districts and Liverpool Street. The beginning was made in the North London Railway (City Trains) Act, 1861, section 45, which stipulated that one train should be run daily each way, before 7 a.m. and after 6 p.m. respectively, between Kingsland Station and Liverpool Street, at fares not exceeding 1*d.* for each journey, or 2*d.* for the return journey.[1] Similarly, in 1864, the Great Eastern Railway, on being granted powers to build a station in Liverpool Street, was called upon to convey workers from and to Edmonton and Walthamstow at return fares of 2*d.* (amounting to about a farthing a mile), before 7 a.m. inwards and after 6 p.m. outwards.[2] The development of the densely populated working-class districts round Edmonton, Walthamstow and also Stratford is thus clearly traceable to the deliberate policy of providing very cheap transport to what were then outlying districts, in order to allay the congestion in the centre of London.

These trains to and from Liverpool Street have become the prototype of workmen's trains in this country, which were introduced as a general institution by the Cheap Trains Act, 1883. The Act abolished the Passenger Duty on all fares of not more than 1*d.* a mile and at the same time required the Railway Companies to run " proper and sufficient workmen's trains . . . for workmen going to . . . and returning from their work . . . between 6 o.c. in the evening and 8 o.c. in the morning as appear to the Board of Trade to be reasonable ".[3] The supervision has now passed to the Ministry of Transport. The tickets issued are return tickets. The time-limit for the journey home has since been relaxed—the companies having termini in London agreed in 1894 to allow holders of workmen's tickets to return any time after noon,[4] but no extension of the morning hour has taken place. The railway companies run a certain number of workmen's trains in the early morning, and any person

[1] *Metropolitan Workingmen's Trains.* A Narrative of the Travelling Tax Abolition Committee, 1893.

[2] J. Blundell Maple, M.P., *Cheap Trains for Workers,* 1891. See also : E. J. Harper, " Statistics of London Traffic " in *Journal of the Royal Statistical Society,* June 1904.

[3] Para. 3 (1*b*) of the Act.

[4] *Modern Railway Working,* Vol. VII, p. 206.

travelling on these trains can do so at the low workmen's fares.[1]

The remission of the passenger tax thus constitutes a contribution towards the lowering of workmen's fares at the costs of the Exchequer. The scale of the benefit derived by the tax remission may be gauged from the fact that in the case of the Great Eastern Railway it amounted, in the 'nineties, to about £70,000 a year.[2]

In fact, the workmen's trains do not cause actual loss to the railways, for the takings usually cover the cost of the train service, though not the overheads.[3] It can be understood that the railways in this country, as profit-making companies, are not financially interested in workmen's traffic. Mr. K. G. Fenelon sums the situation up as follows: "The railways . . . do not make much profit out of workmen's fares, as the charges are so low,[4] but there is a legal obligation on them to provide reasonable facilities."[5]

Nearly one-fifth of all passenger journeys on British railways (in 1938 over 244 million, or 19·8%) are made on workmen's tickets.[6] At first sight, this ratio and the number may seem impressive; however, in Belgium the proportion is nearly one-half of all journeys (47·6% in 1925, for instance); it must also be remembered that a worker travels twice daily, i.e. 600 times in a full working year. On this basis, only 407,300 persons availed themselves of workmen's tickets, in 1938, on the railways of Great Britain (excluding London Transport).

Of the road services, tramways are, like railways, compelled by law to issue reduced workmen's tickets. Where they are run by municipalities, the situation resembles that of the Continental State railways: losses due to low workmen's fares are borne by

[1] In fact, the accommodation provided at the time of demand is often insufficient—see below, p. 61 f. [2] London Reform Union, Pamphlet No. 79 (1899), p. 7.

[3] F. J. C. Pole and J. Milne in *Modern Railway Working*, Vol. VII, pp. 206–7. Cf. further Mr. Frank Pick, *Evidence*—Barlow Commission, p. 360: "On the railways and trams you have workmen's fares at half rate. . . . The trams and trolley-buses are not run at a loss. They do not earn in full the interest on the capital invested in them."

[4] Workmen's fares are on a tapering scale, starting at 2*d.* for a return journey of one mile each way and amounting to one farthing a mile after the twentieth. The average charge on workmen's tickets was 0·53*d.* per mile in 1927 (W. V. Wood and C. E. R. Sherrington, *The Railway Industry of Great Britain, 1927*, Memorandum No. 11, Royal Economic Society, p. 16.)

[5] "British Railways since the War," in *Journal of the Royal Statistical Society*, 1933, p. 394. See also D. N. Chester, *Public Control of Road Passenger Transport*, 1936, p. 208.

[6] *Returns of the Railways of Great Britain*, Summary Table A 1 (*d*), pp. 22 and 26. Season tickets accounted for 31·3% of all journeys.

the ratepayers. On buses workmen's tickets are sometimes issued for reasons of competition with trams. However, in some instances at least, the term "workmen's ticket" has changed its original meaning: the fares are at the level of season tickets rather than so low as to comply with the rates of workmen's tickets imposed by statute.[1]

The number of passenger journeys and the use of the various types of tickets in the London Transport Area is shown by the accompanying table. Workmen's tickets amount to 23·6% of all local journeys on the main railways, and to between 14 and 17% on the Board's railways, trams and trolley-buses. On the buses they are not issued, and as the latter convey half of the

LONDON TRANSPORT AREA

Number of Passenger Journeys (in thousands) according to Means of Transport and Type of Ticket

Year ended 30th June, 1939

Means of Transport	Journeys by type of ticket					
	Total = 100	Ordinary 1–3 class	Season number	Season % of col. 1	Workmen's number	Workmen's % of col. 1
Main Railways [a]	586,554	230,782	217,620	37·1	138,152	23·6
L.P.T.B.:						
Railways	472,665	318,445	75,306	15·9	78,914	16·7
Trams	516,133	432,851	—	—	83,282	16·2
Trolley-buses	570,605	489,169	—	—	81,436	14·3
Buses and Coaches	2,222,695	2,221,056	1,639	0·1	—	—
Grand Total	4,368,652	3,692,303	294,565	6·7	381,784	8·7

[a] Parties to London Passenger Pooling Scheme.

Compiled from L.P.T.B., *Sixth Annual Report*, p. 45.

total passengers in London, the average percentage of workmen's tickets on all services is only 8·7. It must be kept in mind, however, that many bus rides only represent a part of the total journey: the bus is often used as a complementary service to

[1] This can be seen from the fares on country buses in the Oxford region; from Kidlington, e.g., the weekday season ticket to central Oxford costs 15*s.* monthly, while workmen's tickets to Cowley (at the far side of Oxford) amount to 19*s.* 6*d.* (*Time Table of the City of Oxford and District Motor Bus Services*, Edition No. 119, p. 78).

the railway from the home to the station and from the station to the factory or shop. In quite a number of instances the cost of the daily journey is made up by a workman's fare for the train plus an ordinary fare for a bus. Travellers who change buses on their journey to work have to buy two ordinary tickets each way. 60% of all London bus and coach rides were penny rides in 1937, and feeder services doubtless accounted for a high proportion of them. (Barlow Commission, *Evidence*, p. 372.)

Other implications of the workmen's trains will be discussed below in the sections "Hardships of the Daily Journey" and "Measures of Relief".[1]

b Employers' Share in the Cost of Transport

As a rule, the employer does not directly share in the expense of the worker's journey to work. Notwithstanding this fact, employers are definitely concerned with the fares to be borne by their employees, for the obvious reason that they have an influence on the wage-level.

Wages paid by a firm which has all its workers living within walking distance need not take travelling expenses into account. Where, however, labour is recruited at greater distances, the wages have of necessity to be high enough to permit the payment of fares, unless everybody cycles. This consideration must be assumed to influence the various agreements on wages, although in most cases it is not clearly ascertainable. As implied in the term, collective agreements fix wage-rates for a whole industry (or branch of industry) on a national or regional basis; they do not take account of high travelling costs caused by the isolated location of either factories or residential districts. The higher expenses of travelling prevailing in the large towns as compared with smaller places are certainly partly responsible (the other main item being higher rents) for higher wages in the former, particularly in London.[2] According to a *Report on Collective Agreements between Employers and Workers*,[3] some national agreements classify towns and districts; some group them into seven classes of places; some provide higher rates in the Home Counties.

[1] The problems of workmen's tickets from the point of view of the transport undertaking are discussed in *Municipal Trading*, by Herman Finer, 1941, Chapters 15 and 19.

[2] See the juxtaposition of wage-rates in London and the larger provincial towns in the Barlow *Report*, p. 320.

[3] Ministry of Labour; Vol. 1 (1934), p. xvii.

> The principles on which grading proceeds are not usually enumerated in the agreements, but in the building industry . . . some of the matters taken into consideration are indicated in a questionnaire which must be completed before application for grading . . . is entertained. They include . . . transport facilities . . . in the locality.[1]

Thus, except in the building trade, the costs of the journey to work are not perceptibly taken into account in the fixing of wage rates. There are, however, instances of direct contribution by firms towards their employees' transport to work.

Employers' subsidies are either permanent or temporary. The latter usually reflect changes in the demand for labour. Employers undertake to subsidise workers' transport costs when undertakings are started or expanded ; when a factory is transferred to another site ; when a multiple concern reorganises its branches ; when a special scarcity occurs of a particular type of labour. Some examples may illustrate these various cases.

A highly differential system of employers' shouldering the costs of their employees' travelling was adopted during the rapid development in the 1920's in some districts in Germany. It became usual for certain firms to subsidise the fares of their workers on a sliding scale proportional to the distances covered by the daily journeys. This was called *Entfernungszuschlag*. The allowances nearly equalled the total amounts of the fares. Foremost among these firms were the huge plants of the chemical industry, which entered the labour market only after the last war, with works situated in the open country (plants at Leuna and in the Bitterfeld and Wittenberg districts). The employers' federation in the old-established machine industry, on the contrary, explicitly forbade its members to pay such *Entfernungszuschläge*.

In the period of rapid growth of the N. V. Philips Gloelampenfabrieken, Eindhoven, Holland, this firm provided bus services for its workers on a large scale at less than economic costs. Though the passengers paid fares, the services involved a loss to the firm, as the vehicles were idle between starting and stopping

[1] In the building industry, workers have, of course, no fixed workplace, and costs of travelling to the changing sites where building goes on are an item of irregularity which is liable to upset the budgets of the workers concerned, unless an allowance is made for it. There are really two types of allowance in the building industry, first, the grading of wage rates according to the average convenience of travelling in the district, and secondly, compensation to individual workers for exceptionally long journeys.

hours at Philips', except for a small amount of travelling by the employees' families.[1]

"When our works first moved from Edgware to Hatfield," states a large engineering firm, "we provided free transport for such as required it for a period of three to four months." In other cases, the period of adaptation of housing and transport to the new location of a factory, and therefore the need for subsidisation of fares by the employer, takes not months but years.

After a reorganisation, involving the closing-down of some pits, in South Wales, "nearly 300 men and boys from Dowlais found employment at . . . collieries at distances of two and four miles respectively, where they were conveyed by train at the expense of the Dowlais Company." [2]

The shortage of juvenile labour in the western part of London in the mid-1930's led some firms to offer free transport to boys living in East London, either on the L.P.T.B. system or by special buses.

Permanent contributions to the costs of transport by employers range from financial investments for the improvement of railway facilities for the firms' employees to such modest services as the provision of cycle-sheds and works canteens.

It is not uncommon for big works to build, or contribute towards the costs of, stations at the doors of the factory for the convenience of their employees. The Siemenswerke, Berlin-Spandau, with an aggregate personnel of 65,000 in October 1928, undertook an even costlier scheme—the construction of a line connecting Siemensstadt with the existing passenger railway.

Recent years have brought an increase of auxiliary services on the part of firms in connection with their employees' journeys to work. As the bicycle is found by many workers to be a convenient and economic means of transport, firms are faced by a growing demand for orderly storage places near to where the cyclists arrive and depart.[3]

The growing distance between home and workplace has at least partly accounted for the provision of works canteens by a great number of firms. "Canteens are an employees' service,

[1] F. Bakker Schut, *Industrie en Woningbouw*, 1933, p. 91.

[2] Hilda Jennings, in *Second Industrial Survey of South Wales*, Vol. III, page 130.

[3] The price of cycle-sheds is between 22*s*. and 60*s*. per accommodated bicycle (exclusive of extra structural items), and about 10*s*. to 20*s*. per cycle for simple racks.

not regarded as a profit-making concern." [1] While the food costs are usually covered by the takings, the initial capital outlay for the canteen and equipment is borne by the firm. Heating and lighting are mostly paid for by the firm, the wages of the canteen staff in some cases. Formerly, the additional costs of taking the midday meal away from home were borne by the worker. Many had their dinner brought to the factory by their wives or children, continuing the rural custom of carrying the midday meal to the workers in the fields. In one case at least, transport to the factory of dinners prepared in the various homes was organised and appropriate vans put into use. For a certain factory in Baden, the cost of this transport was estimated, before the last war, at 5 to 10 pfennig per meal, and accordingly a minimum cost of 20 marks annually per worker was put down for this item.[2]

c The Cost of the Daily Journey to the Employee

It is the exception to find the employer taking a recognisable share in workers' transport costs—as a rule the costs of the journey are borne by the employee. According to the amount of travelling expense they incur, earners can be classified in three sections : expenses are nil for those who work within walking distance of their homes ; expenses are comparatively low for cyclists ; and expenses vary greatly for users of public means of transport. The available statistical evidence does not allow an assessment to be made of the proportions of the three sections in the country as a whole or in single districts. Yet it is this proportion, together with the amounts of individual fares, which determines the average costs of travelling to work for any given body of earners.

i actual costs

A country-wide survey would be needed to find out the number and distribution of the earners who work within walking distance of their homes and are thus free from travelling expenses. Not even an estimate can be ventured. The statistical material collected for this study is probably weighted on the side of travelling, as attention has been mainly given to instances

[1] *Canteens in Industry*, published by the Industrial Welfare Society, first edition, 1939, p. 29.

[2] Page 65 of the study reviewed on p. 124.

where a fair amount of daily travelling was assumed to take place. The synopsis [1] points to a very great variation in the proportion of earners who walk from door to door; the range is from 71·5% to 1·5% of a firm's employees. At most of the firms under review, however, employees who walk and employees who travel to work are more evenly mixed.

A similar diversity in the incidence of travelling costs prevails amongst the inhabitants of a given place of residence and between various residential places. Of the householders in L.C.C. block dwellings (mostly in central districts) 28% work locally and pay no fares, of the householders in Becontree and in St. Helier, only 9%.

Cycling is a cheap means of transport, and it is used by millions of earners for their daily journey. It is estimated that 8 million bicycles are used during the summer months, and 4 million in the winter season.[2] Not all bicycles serve the journey to work: the figures include those which are used only for the purposes of shopping and pleasure and by children on their way to school. It is clear, however, that without the widespread use of the bicycle the nation's expenses for travelling to work by public conveyances would be considerably higher.

No information is available on the local distribution of bicycles, but it seems safe to say that the countryside and smaller towns benefit to a larger extent from cycling than towns of metropolitan size; and within the conurbations, the outskirts lend themselves better to cycling than the streets in the central districts where traffic congestion is notorious. The relief of travelling costs by the use of the bicycle is therefore presumably higher in small and medium-sized places than in London or Liverpool. The costs of cycling consist mainly in the purchase of a bicycle; the price, in normal times, is from about £4 or somewhat higher if paid in weekly instalments of 2*s.* or 2*s.* 6*d.* spread over a year. After this sum has been paid there remain only the relatively small expenses for repair and maintenance and possibly insurance, and in some cases for special protective clothing. Many people, however, use their cycles during the summer only and join the ranks of fare-payers during the wet season. The problems involved in cycling to work are mainly of

[1] pp. 164/5.

[2] Communication, in 1937, from the British Cycle and Motor Cycle Manufacturers, Ltd.

a non-financial nature and will find their place in the section on the strain of the daily journey.

Fares are the main form of expenses of travelling to work. The total amount of fares paid is no better known than the total number of fare-payers. Various investigations show a wide divergence of fares ranging from a penny to well over 1*s.* a day, that is from 6*d.* to upwards of 6*s.* a week. The accompanying table summarises some of the findings on the range of weekly travelling costs. Whether the level of fares is low, as in the Merseyside sample, or high, as in the case of the L.C.C. cottage estates, there is in most instances a considerable range of travelling costs: in the latter case, over a quarter of the daily travellers paid under 3*s.* a week, while nearly another quarter paid over 6*s.* The average amount per fare-paying miner in the South Wales Coalfield was 2*s.* 3½*d.* weekly; in one district, however, for

Range of Weekly Fares

Summary of Sample Inquiries

Groups of Earners	Total fare-payers under review	Percentage of fare-payers who pay: under 2/-	2/- to 3/-	3/- to 4/6	4/6 to 6/-	6/- and more
Carreras' Employees [a]	779 [c]	37·5		34·0	18·0	10·0
Riverside Workers, Merseyside [b]	371	30·5	52·5	17·0		
		up to 3/-		over 3/- to 6/-		over 6/-
Tenants, L.C.C. [a]						
Cottage Estates	43,477	27·3		48·6		24·1
Block Dwellings	11,294	66·9		28·9		4·2
		up to 2/6	over 2/6 to 4/-	over 4/- to 5/6	over 5/6	
Earners, 4 London Boroughs [a]						
East Ham	365	31·3	37·2	17·4	14·1	
Stepney	483	72·2	23·5	2·5	1·8	
Tottenham	246	52·8	26·8	9·0	11·4	
Willesden	282	57·2	29·4	5·1	8·3	

a For particulars, see Statistical Part, below.
b Source: *Social Survey of Merseyside*, Vol. II., p. 137.
c Including 0·5% unascertained cases.

which particulars are given, the average fares of those who travelled by bus or train were 3s. 7½d.[1]

The differences between various places or districts are increased if the incidence of high fares goes with a high proportion of fare-payers amongst the total earners in one place, while in another place there is a small proportion of fare-payers and these pay only low fares. Such contrast is illustrated by a comparison between a central and an outlying London borough. The sample of Stepney showed well over half of the earners working locally and the remainder averaging a fare of 2s. 2d. weekly. East Ham, on the other hand, had only between a third and a quarter of its earners working in the neighbourhood of their homes, and the 71 % who travelled to work paid a weekly average of 3s. 7½d. in fares.

The outstanding feature of these examples, and indeed of every inquiry into the matter, is the remarkable variation in the burden of fares, and that between various localities as well as amongst the employees of a single firm and amongst the earning inhabitants of a single place or district.

The question arises therefore as to the relative costs of the journey to work: how do the expenses of daily travelling compare with the income of the respective workers, and for what proportion do they account of the total weekly expenditure of the family?

ii RELATIVE COSTS

Fares and Wages.—The information so far available does not reveal a correlation between fares and wages: it has not been found that the amount of fares paid by various persons is proportional to their earnings. Not even a definite upper limit has emerged for the fares in proportion to wages. There must, of course, be a limit beyond which the worker will rather forgo the job than pay the fares, unless the employer is prepared to contribute towards the cost (the question of choosing the cheapest means of transport also appears at this juncture). It is probable that from an examination of more extensive and varied statistical material certain correlations between wages and travelling costs would emerge, though qualified by other factors. Pending further information, the material in hand yields some interesting

[1] Hilda Jennings, loc. cit., p. 38 and p. 107.

results, notably a distinct difference between the position of primary and secondary earners. The self-dependent worker, especially the main bread-winner of the family, cannot spend more on fares than would leave at least a bare subsistence. This limit is not effective with secondary workers; juveniles in particular are not expected to maintain themselves entirely. The restricting influence of transport costs on the choosing of a job is, therefore, of less concern to juveniles than to men, and in fact many boys and girls are found whose fares absorb substantial proportions of their earnings. Even when the fares for juveniles are lower, the rebate is not great. This is made evident by the sample inquiry in four London boroughs.[1] The respective travelling costs of juveniles and adults were 2*s.* 4½*d.* as against 2*s.* 11*d.* in the case of sons, and 2*s.* 3*d.* as against 2*s.* 10*d.* in the case of daughters. Until they reach their late 'teens, however, juveniles earn in peace-time under 20*s.*, many of them only about half this sum. Similarly, information given for miners in the South Wales Coalfield showed that in some cases where fares for adults were 4*s.* a week, boys enjoyed a reduction of only 6*d.*[2]

Many women, too, are secondary earners; their wages, like those of juveniles, serve to increase the combined family income rather than to provide for the total maintenance of the individual earner. Their transport costs need not, therefore, be restricted to a small fraction of the wages, as is the case with the primary earner. This explains why fares are allowed to swallow up high proportions of the gross wage of females. At Carreras' cigarette factory over one-fourth of the reporting girls paid 4*s.* 6*d.* or more per week, and another third paid between 3*s.* and 4*s.* 6*d.*[3] A girl earning £2 might thus spend 10% or more of her wage on transport to work. Housewives, as far as the available material goes, appear to be in a special position: on taking a job, they have a double range of duties and find it, therefore, imperative to avoid lengthy journeys; their average fares are low. (See p. 188.)

But juvenile and women workers by no means regard high fares as negligible. The anxiety felt on this score by committees for juvenile employment in London has already been mentioned.[4]

[1] See p. 188.
[2] *Second Industrial Survey of South Wales*, Vol. III, p. 106.
[3] See p. 171. [4] See p. 22.

The need to keep travelling expenses down is shown by the use of workmen's trains, which are cheap but run only early in the morning. The statistical returns on workmen's tickets do not specify the issue to various types of workers ; however, a visit to a London railway terminus between 8 and 9 a.m. shows numbers of juveniles and of women waiting in the station : although their work does not start before 9 o'clock they are forced to sacrifice as much as an hour each day in order to benefit by the reduced fares on these trains.

Fares as an Item in the Family Budget.—The discussion of the travelling costs of the secondary earners leads to the conclusion that the amount spent on fares must not in the main be related to the wages of the individual earner ; what matters more is the family income, made up of the pooled incomes of the various earning members. The sum of their expenses for the journey to work has to be fitted into the household budget.

Fares occupy an intermediate position between the necessities of life and the dispensable items. Travelling to work is not of such importance as food, clothing and shelter—after all, very many earners still live within walking distance of their workplace and are not, therefore, charged with any expenses under the heading. For a large proportion of the working population, however, daily travelling has become a prerequisite in securing or retaining a job. For them, the net wage available for consumption is the gross wage less the amount of the fares for the daily journey.

There are few items of expenditure in the working-class budget which range from nil to many shillings a week. Drinks and tobacco resemble fares in this respect, but they are semi-luxuries, and spending under these headings is at the discretion of the people themselves. Travelling expenses, on the contrary, are in most cases capable of reduction only within narrow limits by the persons concerned. Another feature which makes travelling costs specially burdensome is the fact that fares have to be paid in cash at the time of travelling, or, in the case of weekly or season tickets, even in advance ; there is no possibility of deferred payment such as can be arranged with the purchase of many household goods, and with the rent in cases of emergency.[1]

[1] Several small money-lending firms were known, before the war, to advance money, at high rates of interest, for the purchase of season tickets.

In view of their importance as an item of domestic expenditure it is astonishing that so little is known about the actual burden on the family budget of the costs of the daily journey. One would, on first thought, look for information to the various calculations of the cost-of-living index, based on investigations of household expenditure. This expectation, however, is disappointed, for reasons deriving from the peculiar nature of the travelling costs which make it difficult, if not impossible, to gain reliable knowledge on this point from sample inquiries into family budgets. This becomes obvious from an analysis of the method and results of the latest Ministry of Labour publication on the subject.

This investigation of "Weekly Expenditure of Working-Class Households in the United Kingdom in 1937–38" [1] has, for the first time, not only specified fares, but subdivided this item (No. 82) so as to show separately the fares for the journey to work.[2] The result obtained by this sample inquiry is an average of 1*s.* 6½*d.* per family, representing 1·8% of the total expenditure of 86*s.* 3*d.* a week.[3] Although it is not impossible that this amount and ratio are fairly correct averages for the whole country, there are reasons for considering them to be understatements.

The main reason is the selection of families according to their local distribution. Painstaking care has been applied to have all regions of the United Kingdom as well as the various occupations and types of family adequately represented. But it is one thing to make the distribution of sample families representative from the point of view of the cost of living in general, and another to do so from the particular point of view of the daily travelling costs. Little is known about the incidence of fares, apart from the fact that they are conspicuously sensitive to location. Within each region there are signal differences, both in the proportion of fare-payers and in the incidence of high fares. This applies to the South Wales Coalfield and to Stockton-on-Tees, as well as to London and Liverpool.

[1] *Ministry of Labour Gazette*, December 1940. Excl. agricultural households.

[2] In the former compilation of the cost-of-living index number, fares had been lumped together with: soap and soda, domestic ironmongery, brushware and pottery, tobacco and cigarettes, and newspapers. All these items together had been assessed an aggregate of 4% of the costs of living; it is clear that not much weight can have been allocated to fares. Source: Ministry of Labour, *The Cost of Living Index Number; Method of Compilation*, 1939.

[3] Plus 8½*d.* for journeys other than to and from work.

Minor points, at least suspect of loading the results towards understating the expenses for daily travelling, are: (*a*) the inclusion in the inquiry of unemployed persons; (*b*) the unspecified proportion of a group of earners many of whom must be assumed to live near their workplace, namely permanent employees of railway companies, local authorities, public utility undertakings and Government departments, with wages or salaries not exceeding £250 per year; railway employees, moreover, travel on free passes—and (*c*) neglect of the costs of cycling.

In view of all these uncertainties it remains, to say the least, an open question whether the collection of 9,000 budgets of working-class families from the whole United Kingdom has hit the true average of daily travelling costs of millions of families.

Professor Bowley and Miss T. Schulz have excluded the fares to and from work from their recent analysis of Working-Class Budgets.[1] In their sample inquiry, which includes towns of various sizes, expenditure for fares between home and workplace was accounted for in 51 out of the total 168 budgets under review and ranged from 6*d*. to 7*s*. 6*d*. per week, averaging at 2*s*. 4*d*. per fare-paying family a week.

The results on the costs of daily travelling obtained from a general inquiry into household budgets have thus proved to be inconclusive.

Another approach proves equally unsatisfactory; it errs on the large side. Based on traffic statistics, an estimate of the travelling costs per family has been made for the London Transport area.[2] The receipts from fares, in 1936–7, amounted to over £41 million. As a population of 9,700,000 was assumed to be served by the London transport system, this meant travelling expenses of £4 6*s*. 2*d*. per person or about £15 per family per year. The average income of the working-class family was taken to be £190 annually, and from these data it was concluded that the total travelling costs were "about 8% of the average income of working-class families". This computation, however, which is being widely used in public discussion, is fallacious. The reasons why

[1] Institute of Statistics, Oxford. *Bulletin* Vol. II, No. 9, p. 4. The authors assume that in some cases fares were not recorded as household expenses but paid out of the pocket-money retained by the earner.

[2] Barlow *Report*, p. 69; Barlow Commission, *Evidence*, p. 358; and *Fourth Annual Report* of the L.P.T.B., pp. 11 and 48.

the result of 8 % is definitely an overstatement are the following. The average travelling costs, which are here related to the income of the poorer section of the population, include, on the one hand, the fares for journeys in which poor families scarcely take part, such as season tickets (including first class) and ordinary first- and second-class tickets on the railways, fares for visits to shops and theatres in the West End and for week-end excursions, including coach rides, further, the fares for local journeys by visitors to London from the Provinces and from abroad. On the other hand, it is ignored that workmen's tickets cheapen the travelling costs of the poor, and that the fares on trams and trolley-buses, which chiefly serve working-class districts, are on a lower basis of charge than the bus fares.[1] Finally, it is not known whether the poor travel as often and over such long distances as better-off people. The proper calculation would be to relate the total average travelling costs to the (unknown) *total* average income of the Greater London families, which would result in a lower percentage taken up by travelling costs; this would, moreover, have to be qualified by the statement that the workers enjoy certain reductions and that the poor make fewer pleasure rides (which are estimated at about one-third of the total journeys).[2]

As general sources fail to produce satisfactory results on the travelling expenses of the family, these expenses must be made the object of special investigations. But most special inquiries have started from the workplace and could not, from the material thus obtained, examine the burden of fares on the whole family unit. Only a few small sample inquiries are so far available for giving some indications of the position.

In the Statistical Part,[3] the aggregate fares for the journey to work have been computed for 923 working-class families with more than one earner each from four London Boroughs. The result, referring to 1930, is an average of 4*s*. 2*d*. per week per family. If we assume, for the sake of argument, that the weekly income per family (with several earners, including juveniles) was £5, the fares amount to 4·2 % of the total income. The average of 4*s*. 2*d*. includes those families who have all members working locally and thus no expenses for daily travelling. The average

[1] Barlow Commission, *Evidence*, p. 360.
[2] Ibid., p. 367.
[3] See p. 155.

weekly fare per family with more than one earner ranges from 2*s.* 8*d.* in Stepney to 6*s.* 8*d.* in East Ham.

In a survey of three Brighton housing estates, 2*s.* to 3*s.* per family per week were found to be the most common amount of transport costs; and "wages are low in Brighton".[1]

On the Liverpool housing estate, Norris Green, the amount of fares for the daily journey was, in 1937, 2*s.* 6*d.* per household per week, despite the fact that nearly half of the earners walked or cycled to work.[2]

On new housing estates where very few are employed within walking distance of their homes, the family's aggregate expenditure for travelling goes up nearly at the same rate as the number of earners. This is clearly demonstrated in a report on the Kingstanding Estate in North Birmingham.[3] Among families with one adult male earner only, the most frequent entry as to costs of transport is 2*s.* to under 3*s.*; among families with two earners, 4*s.* to under 5*s.*; with three earners, 7*s.* to under 8*s.*; in families with four or more earners, the most frequent entry is 10*s.* or more. "2*s.* 6*d.* or 3*s.* *per employed person* per week is a fairly typical subtraction that must be made from the earnings on account of distance from work."

In an investigation of 141 budgets from a new housing estate in Stockton-on-Tees,[4] the average weekly fares were 1*s.* 2*d.* per family. There is a noteworthy absence of correlation between fares for the daily journey and incomes. The households were classified in five groups according to income, which ranged from between 25*s.* and 35*s.* to between 70*s.* and 80*s.* a week. Their respective travelling expenses were: 1*s.* 2¼*d.*, 10*d.*, 1*s.* 4*d.*, 1*s.* 4*d.*, and 1*s.* 3½*d.* In the lowest income group, which comprised about one-third of all families under review, the average fares for the journey to work amounted thus to between 3·39 % and 4·75 % of the total income of the family.

Most of these samples, it is true, refer to new housing estates on the fringe of the towns, where the burden of fares is higher than in the older districts, but in fact a quite considerable proportion of workers are now housed on the outskirts. In Bir-

[1] Marion Fitzgerald, *Rents in Moulsecomb, Brighton*, 1939, p. 39.

[2] *Population Problems of New Estates*, University of Liverpool, 1939, p. 18.

[3] M. S. Souter, E. H. Wilkins and P. Sargant Florence, *Nutrition and Size of Family*, 1942 (the inquiry was made in the Spring of 1939), pp. 44 and 45.

[4] G. C. M. M'Gonigle and J. Kirby, *Poverty and Public Health*, 1936, pp. 198, 201.

mingham, for example, only 18·8% of the working-class population live in the Central Wards, 28·6% in the Middle Ring, and 52·6% in the Outer Ring.[1]

Fares and Rent.—The conclusion from the foregoing sections is that fares are not proportional to wages: neither are the travelling expenses of individual workers taken into account in the determination of wages, nor are the costs of travelling in any way assessable as a fixed percentage of the respective individual earnings or of the pooled incomes of the families. This would mean that earners living within walking distance of their workplace enjoy, in a theoretical sense, a quasi-differential rent to the amount of their colleagues' travelling costs. Or is the non-expenditure on fares (or the payment of low fares) offset by higher costs of housing? It has indeed been widely held that such compensation takes place, in other words that as a rule high fares go with low rents, and vice versa. In the *New Survey of London Life and Labour* this view was voiced as follows:[2] "The cost of travel to and from work commonly bears an inverse relation to the level of house rent, the workman who moves away from the congested centre frequently paying more for travel and less for housing." This opinion is based on the assumption that housing in the town centre and housing on the outskirts are the same thing, so that the location is the only variable in the comparison. On location depends the cost of the building site, and as sites in the centre are more valuable than at a distance, the rent in the suburbs must be correspondingly lower. This theory of a complementary character of fares and rents is borne out by the facts under certain conditions, namely where the comparison concerns dwellings of similar age, size and type. Oxford workers, for instance, taking houses in Kidlington are indemnified for high travelling costs to Oxford by rents which are definitely lower than they would have to pay for similar houses in or nearer Oxford. Less clear is the following case. The London Passenger Transport Board undertook to prove that "as the fare goes up the rent comes down".[3] This statement covers, however, only the new cottage estates; furthermore, merely the

[1] *When We Build Again*, a Bournville Village Trust Research Publication, 1941 (the figures relate to 1938), p. 44.

[2] Volume I, 1930, p. 97.

[3] Barlow Commission *Evidence*, p. 416.

fares of the householders are counted and no allowance made for the additional fares of secondary earners.

In both foregoing instances, the comparison concerns only houses built since 1919. The samples do not, therefore, give a true picture of the actual condition. In fact, the difference between the accommodation available for working-class families in the centre and on the outskirts of large towns is a main feature of the housing development in the last two decades : most houses in the central districts are old, many of the dwellings outworn, without modern amenities and not self-contained. The outlying suburbs and housing estates, on the contrary, consist largely or entirely of comparatively new buildings. In spite of some Jerry-building, the standard of these houses is greatly superior to the old accommodation prevalent in the central districts of the towns. Some of the new houses were furthermore erected in times of high building costs and high rates of interest. The public subsidies given towards reducing their rent have not been sufficient to balance the aggregate increase of costs—hence the higher rents in outlying districts, which at the same time inflict higher travelling costs on their residents.

This state of affairs is confirmed from various sources. The very material of the *New London Survey* bears witness to the fact that, in some cases at least, fares and rents, far from being compensatory, are more or less directly correlative.

The four boroughs included in a sample inquiry [1] show the following weekly averages for rent and for fares respectively :

Boroughs	Average rent (a) *s.*	*d.*	Average fares per family (b) *s.*	*d.*
East Ham	12	6	6	8
Stepney	10	2	2	8
Tottenham	12	4	4	2
Willesden	16	4	4	7½

(*a*) Total Survey Samples ; *New London Survey*, Vol. III, pp. 355, 395, 403 ; and Vol. VI, p. 461.
(*b*) Families with more than one earner—see p. 186.

This table shows that both rents and fares are lowest in Stepney, whilst East Ham combines the second highest rent with the highest fares of the four boroughs reviewed.

Further evidence for the cumulation of fare and rent is given

[1] See p. 155 f.

by various recent investigations of the cost of living in outlying housing estates.

In Stockton-on-Tees, two central areas from which families were moved to a new suburban estate had average weekly rents of 4*s*. 8*d*. "On their transfer to Mount Pleasant the commitments of the translated families were, by reason of higher rentals, increased by 4*s*. 4*d*. per week."[1] In addition fares up to 2*s*. 6*d*. per family per week became unavoidable.

Similar results emerge from a report on the cost of living in three Brighton housing estates: "It is the high rents taken together with other necessary expenditure that create difficulty" . . . "The cost of travelling to work is generally increased on moving to a new estate, unless tenants already possess bicycles."[2]

It is, therefore, "not uncommon to find people continuing to reside in unsatisfactory houses in the vicinity of their work rather than incur the additional expense of travelling costs from a suburb where housing conditions are vastly superior".[3] Correspondingly, there have been complaints that the inhabitants of outlying housing estates suffer from malnutrition: they are at a loss to make both ends meet, as a result of high rent plus high fares.

The conclusions as to the money costs incurred by the worker for the daily journey can be summed up as follows. Fares, the main costs of travelling, are related to the family budget rather than to the individual budget. Fares are of very uneven incidence. This unevenness is wholly borne by the worker, for under present conditions it is neither compensated by differential wages according to travelling expenses, nor counteracted by the rent.

2 *THE HARDSHIPS*

"When a person sells his service as a commodity he has to present himself where these services are to be delivered; he

[1] G. C. M. M'Gonigle and J. Kirby, p. 118.

[2] Marion Fitzgerald, loc. cit. p. 37, seq.—"Between 2*s*. and 3*s*. a week, either on bus fares or hire purchase of cycles, is a common estimate, but a considerable number have transport costs much higher than that, rising as high as 8*s*. in families with several earners," ibid. Dr. M'Gonigle, too, stresses the point that "the capital cost of the bicycle (probably bought on the hire-purchase system) constitutes, for a long time, a drain upon the family income" (loc. cit., p. 201).

[3] G. C. M. M'Gonigle and J. Kirby, loc. cit., p. 202.

cannot send them on like the merchant who sells material goods."[1] The hardships, together with the costs of the journey to work, are directly felt by the travelling public and are often regarded as the main aspect of the whole matter. The hardships of the daily journey present the unambiguous problem of reducing them as far as possible. For this purpose, it is necessary exactly to know their causes, their nature and their magnitude.

On the whole, the prices of the daily journey, paid in the three media of money, time and energy, are parallel and cumulative: long journeys are usually costly and tiresome as well, while a penny ride is unlikely to involve much loss of time or great fatigue. However, there are also features of interchangeability of the three items, that is to say, it is to a certain degree possible to reduce one or two of them at the expense of the third: an employee who can afford a private car for his daily journey reduces the time and probably the strain, as compared with transport by a public conveyance; a man, on the other hand, who does not mind the exertion of cycling to work minimises the costs and often saves time; a person, finally, who puts up with loss of time by waiting can travel by an early workmen's train at a cheap rate. The first device, the private car, has not so far attained practical importance for the journey to work in this country or on the Continent, whilst it plays an outstanding part in the U.S.A. The second method, cycling, is good for fit persons in fair weather, provided the journey is not too long and does not lead through dangerous streets and traffic jams.[2] The fact that workmen's trains are run only up to 8 a.m. means in many cases a serious prolongation of the time spent on the journey.[3]

In view of the cumulative character, as well as of the interchangeability, of the three media in which the journey to work is paid for, it would be convenient to measure them all by the same yard-stick. An interesting attempt has been made by a Dutch writer to express the burdens of loss of time and of the strain of the journey to work in terms of money, so that they

[1] P. Sargant Florence, *Economics of Fatigue and Unrest*, 1924, p. 61.

[2] The problem of making cycling safe has not yet been solved. Separate cycle-tracks, " regarded by many people as the main hope . . . break off at all road junctions, which are just the spots where accidents are most frequent " (H. A. Tripp, *Town Planning and Road Traffic*, 1942, p. 28).

[3] See p. 61 f.

might be compared as "indirect costs" with the direct money costs. But as shown in the Statistical Part,[1] this attempt has not been successful; the present state of knowledge of the effects of daily travelling on well-being and efficiency does not permit of weighing the duration and fatigue of the journey in percentages of wages earned or fares spent. In the absence of a common denominator for the price paid in the various media, loss of time and strain have to be reviewed separately.

a Loss of Time

Travelling time is entirely charged to the employee—collective agreements about the hours of work do not take account of the duration of the journey. On the whole, firms are unconcerned with their employees' whereabouts before clocking in and after knocking off. Working hours have become shorter, but to a certain extent this gain of leisure time has been eaten up by the lengthening of the daily journey. The hours spent on the way between home and workplace do constitute a loss, for they are lost to sleep or recreation, to family life, education or public activities, in short, to the time at the employed person's own disposal. Travelling, even at its best, cannot be counted as leisure, considering the lack of privacy, of ease and of choice of occupation during the journey. Under favourable conditions, it is true, people read or chat in the train or bus; others play cards; some women knit. But these occupations are hardly practicable in the rush-hour traffic, and besides, they are mere stop-gaps. During the journey concentration is as difficult as relaxation; instead, the mind is given to killing time, and thus a mental mechanisation is fostered which is a general danger in the present organisation of life. Some newspapers and the illustrated press doubtless owe much of their popularity to the length and dullness of the daily journey.

In most cases, it is not mere distance which determines the duration of the journeys: the minutes spent on actual train- or bus-rides account sometimes only for a moderate proportion of the total time taken from door to door. For those who use public means of transport the journey consists of various components. It begins with the walk from home to the stop or station where train, bus or tram is boarded.[2] This walk takes a couple of

[1] See p. 126 f. [2] See p. 138.

minutes in central districts, but in the suburbs, where the building density is low, it may require 15 minutes or more to reach the nearest stop or station. There follows the wait for train, bus or tram—a short interval if all goes well, but a long one if the vehicle is late or overcrowded. The time spent in the carriage on actual travelling is the third stage of the journey. Those travellers are fortunate who can catch a through train to their destination—for many the journey involves one or more changes, each of which means another period of waiting. Then there is the walk from the station or stop to the door of the firm's premises. This is the last stage of the journey for employees of small and medium-sized works. In large factories, however, where a number of workshops are spread over an extensive site, each employee clocks in at his department. This procedure adds as much as ten minutes to the journey in some cases.

Even though the total journey is made up of these stages, the sum of their duration in minutes does not give the whole span of time to be reckoned with. If work starts at 8 a.m. and the time which a given employee needs for the journey from door to door is, say, three-quarters of an hour, it is more likely than not that he or she leaves home at 7 a.m., instead of 7.15 as one would expect. The reason is the necessity to be punctual. Firms enforce timekeeping by their employees by regulations of discipline and wage-cuts ; these vary in their strictness but make the fear of being late a very real one.[1] The transport services cannot be expected to work exactly according to schedule, and delay is beyond the passenger's control, especially where the journey involves changing of vehicles. Travellers have, therefore, to allow for possible delay by inserting a number of minutes, up to over a quarter of an hour, into their morning's timetable. The true duration of the journey to work is thus the net duration plus some additional time to provide a buffer, as it were, against the risk of being late.[2]

A grievous loss of time is imposed on many users of workmen's trains by the discontinuation of this cheap service after 8 a.m. For those whose work starts at 9 o'clock the arrival at the terminus before 8 o'clock means a sheer waste of

[1] In one firm at least, the factory doors were closed five minutes after the start of work ; employees arriving late were not admitted and lost the wages for the whole day.

[2] See pp. 135/6.

time. This question will be discussed in the section on measures of relief.

The available information on the actual time taken up by travelling in general and on the incidence of long journeys in particular is scanty. When data are given of the length of the daily journey, this is usually done in terms of mileage.[1] The distance as the crow flies is easy to determine and may be helpful as a preliminary measuring-rod for the purposes of town and country planning and of transport. What matters for the traveller, however, is the time needed for the journey. The difference can be observed at Corby, the new works town of Messrs. Stewarts and Lloyds in Northamptonshire, where "even those living just opposite the works have to go something like a mile and a quarter round to cross the railway and to reach the works entrance".[2]

Some data on the actual duration of the daily journey have been collected for this study and are analysed in the Statistical Part. They show a remarkable variation between different places and districts, different industries and different types of workers. Some firms, mostly of moderate size, continue, as in the old days, to draw the bulk of their labour from the village in which the factory is situated. Long journeys, on the other hand, prevail in the large towns, especially in Greater London, and in such districts in the country where a shifting of industry has occurred, further among the employees of very large plants where thousands of persons are occupied on one site. In such places considerable proportions of employees have journeys of an hour or more, both in the morning and evening.

So far, no general standard has been evolved for the duration of the daily journey—the time of 30 minutes, which some writers stipulate as the maximum, is in many cases impracticable. On the other hand, it is obvious that the drawbacks of lengthy journeys are being increasingly appreciated. The Minister of Labour (Mr. Bevin) wants "to see such staggering of hours . . . that in the majority of cases we get a person from his home to his work in half an hour . . .".[3] Factory Inspectors have for some

[1] For instance, in the table on the workplaces of the tenants of the L.C.C. housing estates, and in the report on conditions in the South Wales Coalfield (see Statistical Part).

[2] G. Boumphrey, "Industry comes to Corby" in *The Listener*, 13th March, 1935, p. 432. See also p. 92.

[3] House of Commons, 27th November, 1940, *Hansard*, col. 306.

years already taken up the question of reasonable duration of the journey in cases of a firm's application for a permit to work two shifts. In Leek, Staffs., for instance, there were difficulties, "as so many women residing in the Potteries were employed"; a condition was therefore imposed "requiring the employers to provide means of transport. The women employed on eight-hour shifts have special buses, and the journey takes 1 to 1¼ hours, whereas the journey without special buses takes 1½ to 2 hours".[1]

Some firms pay attention to the daily journey when recruiting their staff and do not draw from beyond certain distances; at a department store in Liverpool, for instance, "about half an hour is the time allowed to come to business . . . a few people are living at roughly 40 minutes' train journey, but these are the exception".[2] Another pertinent statement is: "When we are engaging employees, we realise that it is very important that they shall be able to get to these works very easily from their home."[3] This attitude may seem sound policy on the part of individual firms, but it must be kept in mind that if it were general, remote districts and outlying housing estates might be entirely cut off from employment. From declining and temporarily depressed industries some instances are recorded where those employees who live at a distance from the works are dismissed first.[4]

b Strain

It is by now generally recognised that many journeys to work impose a strain on the daily traveller, impairing both his personal well-being and his working efficiency. In the words of the Barlow *Report* (p. 91): "Travelling . . . can hardly fail to have adverse effects on health and to result in fatigue (and) loss of energy . . . There can be little doubt, too, that these adverse effects on the workers are reflected in no small measure on their efficiency and output, and, in turn, on the employers' cost of production." As the wording of this statement indicates, the realisation of the ill-effects of daily travelling under prevailing

[1] *Annual Report of the Chief Inspector of Factories and Workshops*, 1928, p. 52

[2] Answer to the Industrial Welfare Society. It is not clear whether the walk to the station and waiting time are included.

[3] Ibid.

[4] *Second Industrial Survey of South Wales*, Vol. III, p. 35; also occurrences in Holland and Germany.

conditions is founded on general impression rather than on exact knowledge. No special investigation of the strain involved in regular travelling has been published in this country. But in the speech already quoted, Mr. Bevin said that " experience has shown that where you can get a person from his home to the factory in half to three-quarters of an hour, as against the hour or two hours that it is taking now in many cases, you increase production by 9 or 10%". It would be very helpful if the material on which this statement is based could be made public, particularly if it makes possible the determination of an exact correlation between the duration of the daily journey and the output. A few tentative inquiries into this matter have been made in Germany.[1]

The scarcity of information is largely explained by the fact that, unlike costs and loss of time, the strain cannot be assessed in definite units. Fares are measurable in shillings and pence, duration in hours and minutes; but the strain has to be gauged by description of the inconveniences inflicted on the travellers and by the effects on health and efficiency. Attempts (both in previous and in the present investigations) at an accurate assessment of the strain will be discussed in Part B of this study.[2] The results so far obtained do not lend themselves to generalisation, but rather serve to shed light on the complexity of the problem and to indicate the lines which further research might follow. Inquiry into the strain of daily travelling must consider four points: causes of the strain; types of persons affected; different forms of the harm suffered; and effects on efficiency.

The mere duration of the journey is a strain in itself. An hour's travelling both morning and evening, added to a working day of 8½ hours[3] and a midday break of an hour, makes a total of eleven and a half hours away from home. It is obvious that this has a direct bearing on the times and types of meals and on the opportunity for sport and exercise in the open air. Lengthy journeys added to a twelve-hour shift curtail even the hours of sleep. The question is whether a limit can be set, from the point

[1] Friedrich Ritzmann, *Einkommens—und Wohnverhältnisse der Arbeiter der Maschinenfabrik Gritzner A. G. in Durlach*, 1914.—W. v. Drigalski a.o., *Arbeit und Wohnung*, 1931.—Charlotte Grabe, *Der Einfluss der Pendelwanderung auf die Arbeitnehmer*, 1926.—L. Preller " Die Entfernung vom Wohnort zur Arbeitsstätte " in *Reichsarbeitsblatt*, 1925, No. 24.

[2] See p. 124 seq., and p. 139 seq.

[3] Including ½-hour to make up for the free Saturday afternoon.

of view of industrial health, to the duration of the daily journey, either generally or in conjunction with the length of the working day proper.

The circumstances of the daily journey by public means of transport need detailed examination, in order to apportion the responsibility for strain amongst the various features. These are : first, the inconveniences of the various stages of the journey, walk to and from the station, waiting, queueing, changing vehicles ; secondly, the condition in the conveyances, viz., standing sandwiched in overcrowded compartments and the stuffy atmosphere which tires and fosters contagion ; further, superimposed on all other items, a constant hurry and nervous tension due to the fear of being late ; and lastly, equally influencing each phase of the journey, bad weather.

Bad weather is also the chief enemy of cycling in this country ; during the winter months the number of persons who cycle to work drops sharply. The London Passenger Transport Board reports that " inclement weather aggravates the difficulties of meeting the peak flow to and from the factories, as many users of pedal cycles then prefer to travel in greater comfort in the Board's vehicles ".[1] Similar reports, from other parts of the country, were made by member firms of the Industrial Welfare Society. On the other hand, this method of transport is free from most of the drawbacks of public transportation. Not the least advantage is that the cyclist is, to some degree, out at his own time ; one minute lost does not mean for him the missing of train or bus. The Continental investigations pointed to the adverse effects of the physical exertion involved in cycling ; [2] but this was largely ascribed to a hilly countryside and referred, in the years of the German inflation, partly to undernourished workers. Nevertheless, the fatigue from cycling is not negligible. This is indicated by the fact that cycling to work finds comparatively little favour with female workers, in spite of this being often not only the cheapest but also the quickest method of transport.[3] Girls have another reason, however, which may be at least as effective in keeping them from cycling, namely consideration for their clothes. The difference in dress habits which still, if in diminishing degree,

[1] L.P.T.B., *Fourth Annual Report*, p. 26.

[2] These remarks also apply to lengthy walking, though probably not to the same degree as to cycling.

[3] See the findings at Achille Serre Ltd., p. 143.

distinguishes the " black-coated " from the manual workman is much less marked among females. The factory girl's apparel is essentially of the same kind as that of the shop assistant or clerk : it is townwear, not really suitable as a cycling outfit.

It must be assumed that the strain of travelling is felt unevenly by different types of persons ; separate examination is required of males and females, juveniles and elderly workers, office employees and those with non-sedentary occupations.

Too little is known about the various forms of the harm done ; it may result in specific complaints of the respiratory or digestive systems, but it may also make itself felt as general tiredness and nervousness. " Vastly more days are lost from vague, ill-defined but no doubt very real, disability due to ennui than from all the recognised industrial diseases together." [1] How much of this disability and ennui is caused by the strain of the daily journey?

To examine these complaints is actually a matter for medical research ; but in some cases it is possible to get information from a perusal of firm's personnel records on absenteeism and of the pay-roll. If the length and circumstances of the daily journey have an influence on health and efficiency, this should be traceable in the respective employees' proneness to sickness and accidents and in the amount of wages earned by them on piece rates. The relevant German investigations found indeed a definite correlation between the length and discomfort of the journey and absenteeism due to sickness, also a similar correlation between the hardships of the journey and accident rates, and finally an inverse correlation between the travelling hardships and the wages earned.[2] But these inquiries have been in the nature of experiment rather than of final establishment—the matter wants further illumination.[3]

Only after the various disadvantages of the daily journey are assigned to their respective causes and to the persons affected can proper steps be taken to reduce the more serious harm done by the journey to work.

[1] *Annual Report of the Chief Inspector of Factories and Workshops*, 1931, p. 75.

[2] See p. 124 f.

[3] Among the statistical data collected for this study, the material of Carreras' cigarette factory had seemed to make possible an examination of the relation between the absenteeism and the travel burden of the individual employees. However, the number of cases which could be included in the inquiry proved too small for a statistical analysis of this kind, and the results were inconclusive. See p. 140.

3 *MEASURES OF RELIEF*[1]

The price paid for the journey to work in costs and hardships is high; to many writers it seems indeed so high as to justify the demand that daily travel should be abolished. Yet, as has been shown in the preceding chapter, some of the journeys fulfil wholesome functions in the modern social and economic system, and others are caused by hard facts of a topographical nature (namely the existing location of industrial and residential districts respectively) which may often be neither inevitable nor useful, but which cannot now be undone for many years to come. Travelling to work by masses of earners has, therefore, to be accepted as unavoidable for the time being. As no panacea can be found for the complex grievances, it is all the more important, and fortunately possible, to some degree to relieve the various complaints. Sometimes comparatively small measures have a considerable effect.

a Measures by Firms

The Five-day Week.—The five-day week is a helpful device: it reduces by a full sixth the number of journeys and with it the cost, loss of time and strain involved. To minimise the employees' travelling is one of the main considerations prompting firms to introduce the five-day week. The five-day week "is more often found in cases where workers have to travel long distances to their work than where they live around the factory".[2] Instances are given of a firm which had moved from South London to the Reading district: "the five-day system was mainly started because many of the original workers continued to reside in London, and so much time was taken in travelling"; and of cases in Birmingham where "in factories in which labour is drawn from the Black Country it is specially convenient since it is not necessary to make the long journey on Saturdays".[3] The working hours lost through not opening on Saturday morning are usually added to the five remaining days. As shown by the following case, this can lead even to a two-fold saving of fares:

[1] See the author's articles "The Journey to Work" in *Industrial Welfare*, July 1936; January 1938; and April 1940.

[2] *Annual Report of the Chief Inspector of Factories and Workshops*, 1929, p. 52.

[3] Ibid., 1927, p. 57.

> The main reason for adoption (of the five-day week) in the wholesale dressmaking trade in the West End of London appears to be one of location. The factories are situated in an area which is no longer residential, and workers are drawn from long distances and are unwilling to make the journey for a few hours' work on Saturday morning. . . . 9 a.m. was the usual starting time before the advent of the five-day week, when it was changed to 8.30 a.m., and incidentally allowed workers to make use of workmen's tickets.[1]

For factories, the short hours worked on the Saturday make this day the obvious choice for being dropped as a working day. Retail shops, on the other hand, might consider adopting the practice of the big department stores in Paris : in view of the fact that housewives are busy at home after the week-end they used to keep the shops closed on Monday mornings. If different business branches could agree to close on different days, this organisation of the five-day week would not only reduce the total number of journeys but, being a "staggering" of working days, would also relieve the overcrowding of vehicles and the traffic congestion on each day. Whether the reform can be carried so far or not, there is no doubt that from the point of view of lessening the costs and hardships of the journey to work, the five-day system is strongly to be advocated.

"Staggering" of Working Hours.—Much interest has recently been focused on the "staggering" of working hours. While this device does not reduce the number of journeys nor the amount of fares, it relieves the rush-hour traffic. A great deal of the strain of the journey to work is due to the fact that many thousands hurry to their work at the same time. If the cataract of workers arriving at the conventional time can be regulated into an even flow over an extended period, many of the traffic difficulties will be eliminated. Less crowding within the vehicles and less waiting, queueing and straphanging will result. There will also be less congestion in the streets and consequently a shortening of the time taken on the journey.

It must not be overlooked, however, that "staggering" involves considerable difficulties ; it is a problem of business organisation, and it also requires very careful dovetailing of the transport services. An object-lesson, both of the achievements and limitations of the scheme, is the introduction and operation of a staggering scheme at Austin's, Longbridge, which is described

[1] *Annual Report of the Chief Inspector of Factories and Workshops*, 1930, p. 62.

in the Statistical Part. Similar schemes in other countries seem to confirm that "staggering" is a suitable device for very big works or groups of factories, where numerous employees are concentrated on the outskirts of a town.[1]

Some years ago the Siemenswerke in Berlin-Spandau set the times for starting work at intervals of ten minutes for different units. At the Ford Motor Works in Detroit, U.S.A., the "stagger" extends over two hours. In the administrative centre of Washington, "staggering" was necessitated not so much by the crowding of public conveyances as by the congestion of the streets by private cars, because so many of the civil servants arrive in their own cars.

In business and shopping centres, on the other hand, where the conflux consists of the employees of a number of firms, often belonging to the same trade, more intricate problems arise which make it doubtful whether the scheme is practicable at all. Agreement on the starting and stopping times of these firms is necessary; but can an Oxford Street shop, for instance, be expected to close half or even a quarter of an hour earlier than his next-door competitor? Would "staggered" opening and closing times of the various firms be in the interest of the buying public, considering that it reduces the opportunity of selecting the wanted articles from amongst the stocks of several firms? The inconvenience to the customers would be increased if, in order to do justice to the shops, a system of monthly rotation as to the times were introduced.

Some degree of staggering already exists through the fact that large towns, particularly London, have several centres, demanding journeys of different lengths from the main residential districts.

In one important respect, the development during recent decades has brought a compression rather than a staggering of the period when work starts. Formerly, factories began work earlier in the morning than shops and offices; with the shortening of working hours in industry, however, the starting times of manual workers have become later and thus drawn nearer to those of clerical and commercial employees. This fact is of consequence for the appreciation of the Railway Companies'

[1] J. E. Cowderoy describes staggering of hours for outlying factory groups in war-time London in "Public Co-operation in Transport", *Journal of the Institute of Transport*, January 1943.

policy of operating workmen's trains only up to 8 a.m., as will be discussed in a subsequent paragraph.

Other Measures.—It does not seem unlikely that more exact investigations than have been made so far will confirm the harm caused by the fear of being late, both in adding to the time allowed for the journey and as a major cause of nervous strain. In that case, the remedy will have to be a relaxation of time-keeping rules, although these are at present thought necessary in the interest of output and discipline. There are certainly cases in which late-comers disturb the process of production; but not all earners are employed at conveyor-belts making a whole gang dependent on each other, nor are they all members of a team getting group wages. Experience in the present war shows that rigid time-keeping regulations can in many cases be relaxed without serious harm to efficiency. In this as in other respects, good management should not find it impossible to reconcile the requirements of factory, shop or office with the human needs of the employees.

A modest contribution towards shortening the daily journey can be made by firms which have branches situated in various districts of a town, by transferring employees to branches which are nearer to their homes. Such a transfer was carried out some years ago by the municipal Gas Works in Berlin, and it has been practised by Messrs. Marks & Spencer in London. Where it can be applied, this method reduces the length of the daily journey together with the cost and the strain, but it is limited in its possibilities. Employees are transferred only in cases of very long journeys, because a frequent exchange of personnel would be harmful to the team-work in the branches. A strict system of local recruiting is not possible with the present location of residential and business districts, and the advantages of the mobility of labour in a large town would thereby be lost.

As already mentioned, in large factories where a number of workshops are spread over an extensive site, each employee clocks in at his department, which may be as much as ten minutes' walk from the gate. Where access to the works, whether by road or by rail, is from several directions, help can be given to the employees by providing several entrances, while for cyclists further time may be saved by placing cycle sheds at convenient points within the factory grounds.

Within the working day, rest-pauses and the supply of good meals in the works canteen do much to counteract the strain of the journey as well as the normal fatigue of the working hours.[1]

b Measures in the Sphere of Transport

Every technical improvement in the transport services reduces the loss of time and the discomfort of the daily journey. Such measures are the running of through trains, of more frequent trains, of longer trains, and similar devices in road traffic. It is clear, however, that in the more serious cases of overcrowding and congestion such as prevail in the large towns, and notably in London, the success of these measures is relatively slight.

Workmen's Trains.—Workmen's trains, which have been repeatedly dealt with in this chapter, must be discussed here from a particular viewpoint. On the one hand, their *raison d'être* is to relieve the financial burden of the fares ; on the other hand, severe grievances are inflicted on many users of these trains in strain and loss of time. The questions are whether and how the hardships can be mitigated.

It will clear the issue to compare present conditions with those in the early times of the operation of workmen's trains. In the 1880's and '90's, workmen's trains from Walthamstow and Edmonton to Liverpool Street were run only up to 6.47 a.m. " The Railway Companies have somewhat underestimated the evils of a wage-earner being landed in London one or two hours or more before his work begins ", was a judgement in the Court of the Railway and Canal Commission in the Matter of an Inquiry under the Cheap Trains Act, 1883.[2] The situation has not materially changed in the last 40 or 50 years : in 1936, the medical officer of Harrods, London, for instance, made the same complaint about " the increase of the working day by members of the staff (especially juniors) who make use of the workmen's tickets to save expense, though not obliged to travel at the early hour required because of their starting work ".

Workmen's trains, it is true, are now being run up to 8 a.m. (arrival at the terminus), but the number of trains run shortly before 8 o'clock is quite inadequate to cope with the demand, so that many persons have to go by earlier trains ; in addition,

[1] See pp. 35 and 69.
[2] Report from Select Committee on Workmen's Trains, 297, 1903, pp. 44/5.

hours of work, in many London firms at least, do not begin before 9 a.m.; the time-lag between arrival at the station and clocking-in time thus remains. Severe crowding in the workmen's trains and the ensuing discomfort is the other grievance. In this respect, an ironical development has taken place: conditions have become equal to those of ordinary business traffic—not through improvement in the service of the workmen's trains, but through deterioration of the ordinary train service during the rush hours. Overcrowding of the vehicles (and congestion of the streets) before the opening and after the closing of shops and offices is so intense as to be little better than that of workmen's trains. The overcrowding of workmen's trains has thus largely become part of the general problem of the traffic peak. An extension of the operation of workmen's trains beyond 8 a.m., while affording some relief, would on the whole narrow the period of the rush hours and thus intensify the pressure.

This is demonstrated by two diagrams showing the London peak traffic on the District Line (Whitechapel Station) and on the Morden-Edgware Line (Oval Station) respectively.[1] At Whitechapel, the Westbound loadings of the trains are about 12,500 at 8 a.m. (per half-hour), 8,000 at 8.30, and 10,000 at 9 a.m., while the available seats are about 4,000. This means that even in the interval between 8 o'clock, when the workmen's tickets lose their validity, and 9 o'clock, when the ordinary traffic is at its highest, the train loading is twice the seating capacity. On the Morden-Edgware Line, the drop between the two peaks (both about 12,000) is sharper, viz. to under 7,000, while the number of seats is between 4,000 and 5,000. In this case, a half-hour extension of the availability of workmen's tickets would bring some relief (moreover, the seats provided before 8 o'clock might be brought to the full capacity reached for the ordinary traffic an hour later). In outlying factory districts, on the other hand, heavy workmen's traffic is compressed into half an hour, with little ordinary business traffic to follow. In such areas, an extension of the workmen's tickets period beyond 8 a.m. would, from the service point of view, be possible and would facilitate the staggering of working hours. But the main problem is inward travel.

The situation needs consideration from a wider angle. Cir-

[1] L.P.T.B., *Third Annual Report*, Frontispiece.

cumstances with a direct bearing on the operation of workmen's trains have changed so considerably as to call for a reorganisation, or at least for an amendment, of the existing regulations for workmen's tickets. From the beginning up to this day, the reduction of fares by workmen's tickets has been intended to assist low-paid workers, while the transport undertakings were and are to earn the revenue from the fares of the ordinary business traffic. Formerly, when the starting time of factories was distinctly separate from those of offices and shops, the discontinuation of workmen's trains after 8 a.m. served as a criterion of bona-fide workers. As shown above, however, the two periods of traffic have drawn closer together. Furthermore, it is now not always factory operatives who earn the lowest wages, thus most needing cheap fares, but rather juvenile and other low-paid employees of business houses which do not open before 9 a.m. Modern conception of the workers' welfare, however, regards hanging about the station as highly objectionable.

Thus, the intended beneficiaries of cheap tickets cannot now be classified by either time of transportation or rough grouping of occupations. A more specific selection has become necessary, in order to ensure that those—and only those—who need it, get extra cheap transport without sacrificing an hour or more of their morning sleep. Under the present system it is not possible to " pick and choose the people to push up " into the later period, at reduced fares.[1] But such discrimination is just the method which would serve the purpose. In Belgium, as described in the Statistical Part, every buyer of an *abonnement d'ouvrier* has to produce a document showing him to be a bona-fide worker under the terms of the workmen's trains act. In this country, it does not seem necessary to subject the bulk of the users of workmen's tickets to such a scheme, which, moreover, involves considerable administrative work. It should be possible to find an intermediate solution : up to 8 a.m., everybody might use workmen's tickets, as at present, while those who are to be entitled to reduced fares after that hour would need certificates showing their claim ; new regulations would have to fix the categories of eligible persons, mainly according to the amount of wages earned. Such a procedure would only be an expansion of some existing schemes. On the railways, for instance, season

[1] Frank Pick, Barlow Commission *Evidence*, p. 417.

tickets for periods from a month upwards are issued at half rates to workers under 18 years who earn no more than 18*s*. per week.[1]

The principle of a flat rate, i.e. uniform fares for local journeys of any length, is applied in some large American and Continental towns, but has, on the whole, found no place in the transport system of this country. From the point of view of the worker's journey the main considerations are (1) Persons with long daily journeys have their fares appreciably reduced under a flat-rate scheme. (2) Unless the maximum distances over which the flat rate works are fairly short, a flat rate is bound to depend on public subsidies, because traffic over short distances would not bear the increases of fares implied in an equalisation. (3) Flat rates are a temptation to spread the conurbations further and thus to increase travelling and aggravate the loss of time and the strain suffered.

Encouragement of Private Transport.—The advantages of cycling to work, both to the individuals concerned and as a relief of the pressure on public conveyances, make encouragement of the use of bicycles the right policy. This is increasingly recognised by employers who provide cycle-sheds at suitable points within the works premises and facilitate the purchase of bicycles by their employees, for instance by supplying them at wholesale prices. During the winter months, many workers cycle in fair weather only and go by public conveyance when it is wet; the transport agencies are usually expected to cope with this bad-weather demand for additional seating capacity.[2]

It is to be expected that after the war the ownership of private motor-cars will expand. The relief thereby afforded to public transport depends not only on the number of cars thus used but also on the number of persons carried by each car. Up to the present war, it was an offence for several people to contribute to the costs of a car which they used in common unless they bought a Public Service Vehicle Licence and a Road Fund

[1] As a war-time measure, this concession has been extended to weekly tickets, and the wage limit has been raised to 25*s*., since 1st December, 1940.

[2] In view of this habit, an instance like the following cannot be regarded as serving the general interest: In the case of a certain firm the employees who cycle in fair weather cannot avail themselves of the most convenient public means of transport, when they wish to do so; bus services, in this instance, cannot be sufficiently augmented to take all would-be passengers in time for clocking-in. It happened that regular passengers were left behind on rainy days, as "fair-weather cyclists" had boarded the bus before them. Thereafter, by arrangement between the firm and the bus company, only the holders of weekly tickets were admitted to

Licence.[1] This regulation was bound to put a severe check on the sharing of cars by workers, although there is evidence that to some extent the ruling was circumvented. It is to be hoped that the emergency measures which, in the earlier part of the war, encouraged the sharing of cars in order to save petrol, on the one hand, and to ease the pressure of the rush-hour traffic, on the other hand, will be continued, in some form or other, in peace time. A lesson can be learned from conditions in the United States, where the ownership of cars is widespread among all classes of the population, unhampered by the imposition of special licences. Several American investigations illustrate the frequency of the car-sharing habit in that country. In 1936, a special Workers' Transportation Count was taken of the 30,000 employees of four automobile (and accessory) factories in Flint (Michigan) ; [2] it showed that two-thirds of the total reached the workplace by private car, the greater part of them sharing it with at least one other, but more often with two or more persons. 53·9 % of the workers spent 15 minutes or less on their journey, and a further 37·5 %, 20 to 30 minutes ; over nine-tenths, therefore, had not more than half-an-hour's ride. This would have been impossible if two-thirds of the workers had not travelled by car. A small sample inquiry in several places in another American State [3] showed that only in a minority of cases was the car shared by members of the same household. It was non-relatives who were usually picked up at their own houses, each passenger being carried by private car from door to door.

The measures of relief which can be applied in the spheres of factory organisation and transport service are all of relatively

the buses. The Traffic Commissioners have sanctioned this regulation. The bus might, of course, make a second journey to pick up the remainder, but they would be late for work. This firm has not found it possible (as others have done) to relax the time-keeping rules, but expects the fair-weather cyclists to travel by train in bad weather, which takes longer than by bus and means a walk in the rain from the station to the works. Such a policy obviously discourages cycling to work. The firm subsidizes the bus service and, in order to make it pay, deliberately discourages cycling to work by the employees.

[1] In May 1939, five men were fined at Cowbridge (Glamorgan) Police Court for thus "acting in direct competition with the public service vehicle system". *The Times*, 17th May, 1939.

[2] Unpublished material, obtained in 1937 through courtesy of the American Association of Planning Officials.

[3] Massachusetts. I am indebted to Professor Carle Zimmerman, Harvard, for this information.

modest importance. The major solution must be sought in town planning. The urban layout as recommended from the point of view of the daily journey will be discussed in Chapter V. It will there be shown that the burden of daily travelling can be lessened in two ways: the number as well as the length and strain of journeys to work can be reduced.

Chapter IV

THE DAILY JOURNEY AND COMMUNITY LIFE

I Severance of Dwelling Place and Workplace

SO far, the discussion has been focused on the journey to work and its immediate implications. Beyond these, however, the daily journey also has important bearings on the broader aspects of community life. In the following chapters, therefore, the journey itself will be left out of account and consideration given to the problems arising out of the severance between dwelling and workplace and to the implications for town planning.

1 *THE EMERGENCE OF SPECIALISED TYPES OF PLACES*

As a result of the daily travelling to work of masses of earners, " for localities situated within a region of highly organised industry, separate account must be taken of both a night and a day population, the two often differing widely from each other in number and constitution ".[1] According to this " new factor in demography ", four types of places can be distinguished.

(1) Places which are, as in old times, self-sufficient labour markets, that is to say, where the residents, and residents only, work in the locality.

(2) Places whose night and day populations are approximately the same in number, but different in constitution. This occurs where the daily influx and outflow of workers are numerically equal. Such a state of affairs makes the number of people affiliated to the place greater than the number of its inhabitants, for both residents who work elsewhere and persons coming in to work from outside share the town's life in certain

[1] Census of England and Wales, 1921, *Workplaces in London and five Home Counties*, p. iii.

respects.[1] On the other hand, the entire civic interests of neither group are centred in the one place. This kind of place occupies an intermediate position between the first-mentioned and the following types.

(3) Places where the night population considerably outnumbers the day population, i.e. residential places or " dormitories ".

(4) Places where, on the contrary, the day population preponderates ; such are distributive and commercial centres and factory districts. (No term has so far been coined to denote this counterpart of the dormitory.)

The emergence of places of the two last-named types has been a conspicuous feature of urban development in recent decades. Dormitories and exclusive workplaces are complementary, each representing part of the former comprehensive locality. Both types of place come into being either by the transformation of an old locality or by the creation of a new specialised place. These differences of age and origin produce localities of different character.

a Workplaces

An outstanding example of the " divorce between residence and workplace ", the City of London, has already been mentioned. Many other business centres in London and other large towns show similar though less pronounced conditions. Such urban districts undergo an atrophy of the elements pertaining to private life ; in fact, " day " population is only a working-day population, comprising earners only. The significant feature of the transformation of a compound place or district into a business centre is a growing specialisation of its character and functions in accordance with the demands of the breadwinners. Churches are empty on Sundays from the lack of residents in such a district ; the same happens to schools in the absence of a child population ; housewives are almost non-existent—hence the disappearance of shops for the everyday needs of a family. Instead, there is a notable selection of shops

[1] This is illustrated, e.g., by the regulations of the Oxford City Libraries (Abstract from the *Rules of the Lending Department*, Para. 1) : " Books may be borrowed from the Library by all persons . . . who are resident in, employed in, or attend schools in, the City of Oxford."

—on the one hand subsidiaries to the main business carried on, such as stationery, contrivances for window-dressing, or surgical instruments, on the other hand supply of certain consumers' goods and services, shops for men's attire, barbers, tobacconists, perhaps a few bookshops, and gardening and sports shops.

An interesting effect on the life of districts of the distance between home and work is the growth of a catering industry. Establishments catering for the lunches of masses of earners have developed mushroom-like in business centres of all kinds during the last few decades. To have regular meals in restaurants—whether it is a hot two-course meal or a sandwich-snack—is a great alteration, sociological as well as economic, in the habits of the working population; especially for females this is a rather revolutionary innovation.

In many parts of the country, however, working-class households have not quite parted with the custom that the housewife, from the sum handed to her by the earners, has to provide all meals for the family; earners take to the workplace sandwiches or cooked food for warming up. This tradition of the working-class budget has been described as holding back the development of works canteens: the meal in the canteen, even if cheaper or better than the food brought from the worker's home, would have to be paid for from his pocket-money, while the weekly sum handed to his wife or mother covers the supply of a lunch to be taken to the workplace. The habit of taking the midday meal to the workplace has, on the other hand, changed the kind of food consumed by urban workers. One instance is "cubes of condensed beef essences. Now that these are obtainable in small one-person packings, they are much used instead of cold tea or beer by those members of a family who go out to work and take their dinner with them".[1] The war has changed the situation, and the custom of having a hot midday meal in the canteen will probably not disappear when peace comes.

The business centre shows the shrivelling of a complete community with all the many-sided functions of a mature town into a locality where only trade is carried on. In contrast to the older districts which undergo a gradual modification, Trading Estates are a new form of lay-out, meant from the start to concentrate exclusively on industrial activities. "Trading Estates

[1] *New Survey of London Life and Labour*, Vol. VI, p. 306.

set out to offer facilities and attractions to industries seeking location, but provision is not made by the estate companies, except to a small extent, for housing the workers employed . . . on the estate." [1] In Trafford Park, for instance, where approximately 50,000 workers are employed by some 200 firms, only about 700 houses have been built for "the key men who have to be within call of the factories at all hours".[2] Similarly, about 100 houses have been provided on the Slough Trading Estate which has a total of 28,500 workpeople.[3] The success of these two private trading estates stimulated the Commissioners for the Special Areas to establish trading estates in the Depressed Areas. As these estates were started with a view to grappling with the severe unemployment in these areas, they are intended to be located within easy reach of populous residential districts. But they were designed as units by themselves, not as parts of local communities, in other words, as exclusive workplaces. Only in the garden cities, industrial areas were planned in conjunction with residential areas.

Some very large single works where many thousands of workers are employed, or smaller factories in an isolated position, form workplaces of their own. An instance of the first kind is Longbridge, seven miles from the centre of Birmingham,[4] where the Austin Motor Company employs 20,000 persons, while the number of inhabitants is insignificant. An example of the second type is the British Bata Shoe Company, a recent foundation at East Tilbury, Essex.

b Dwelling Places

Purely or predominantly residential places have come into being chiefly by the creation of new dormitory suburbs. In addition, there is the transformation of older villages and small towns which now serve as dormitories for the big labour markets.

The new housing estates are of particular interest. Four million, i.e., about one-third of all houses existing in England and Wales, have been built since the last war, the majority of them either by way of peripheral expansion of the built-up area or in

[1] Barlow *Report*, p. 129. [2] Ibid., p. 283.

[3] Ibid., pp. 284–5. There has, however been some recent housing development near the trading estate.

[4] Longbridge is, however, within the boundaries of Birmingham.

new outlying areas. The size of these dormitories varies from a few hundred inhabitants to the 120,000 of the L.C.C. estate, Becontree. As implied in the name "dormitory", the bulk of the active population is absent during day time. From the outset, the housing estates cover only the residential sphere of life. More will have to be said on the dormitories in the following section.

A good many small towns and villages have become dormitories for large towns. The process is composed of two different developments. Firstly, some of the long-established residents lose or give up their local employment and take jobs in the greater labour market. This happened particularly during a period after the last war when house building lagged behind the expansion of industry in London and other large centres whilst agriculture and local industries declined. It means that the night population of the smaller place remains unchanged in number and composition, but its day population decreases. The second way in which the relation between night and day populations has been altered is the following: while the old place continues to provide employment for the same inhabitants as before, masses of new houses are built within the boundary of the locality for new inhabitants who work in the large town. These dormitory excrescences often outnumber the old self-contained settlement and turn the place, numerically at least, into a dormitory. Where the newcomers are overwhelming in number, as for instance in Kidlington near Oxford, the old village almost entirely loses its character. Abingdon, to take another example in the Oxford region, has equally developed considerable dormitory adjuncts and is now largely a dormitory; but at the same time the old country town has retained much of its own life.

In practice, the two developments are mixed up, and it would be difficult to disentangle them in most of the formerly independent localities which have been drawn into the orbit of a large city. The dormitory quality of various residential areas attains different degrees. Where the dormitory character is imposed on the local life of a self-sufficient locality the degree may be comparatively low; in an isolated new housing estate it is 100 per cent.

2 *THE QUESTION OF CITIZENSHIP UNDER THE CONDITIONS OF SEVERANCE*

The problem of the "severance" of dwelling places and workplaces is twofold. Firstly, all commuters, to use the American term, have their allegiance split, as a matter of definition ; each of the two places concerned may, however, be a comprehensive locality, on traditional lines. Secondly, dormitories and exclusive workplaces are both fragmentary places, each concentrating upon one side of life. The question, from the point of view of the individual, is, therefore, where and how a man can be a citizen in these circumstances ; and with regard to the localities, how the two specialised places can be imbued with a community spirit.

The emergence of a dualism of places for private life and for work respectively is not an isolated phenomenon. In general we find that social development has been accompanied by a relaxation of bonds binding men to a narrow circle, with the result that they can fulfil their various interests in different appropriate circles. Is the severance of the places of domicile and of work not in line with this aspect of social progress ? The old conception of citizenship was perhaps very limited when it implied that a man played all his parts on one stage. From this it would follow that the polarity of places may be well suited to the differentiation of modern life ; that workplaces may exhibit a peculiar and valuable spirit of their own, and dormitories develop a clear consciousness of the needs of another side of social life—in other words, the severance may enable citizenship to be more efficient, though less visible, than formerly, by introducing specialisation in the ways of living.

How does the actual state of affairs compare with this picture of a possible and salutary development ? So far, the hopes for a better and fuller life in the new arrangement of places have not quite materialised. There is a consensus of opinion that as a social achievement, the development of the inter-war dormitories (and, it should be added, of many older residential suburbs) has been disappointing. On the whole, housing estates have not displayed such unifying force as to attain a communal personality of their own and to instil in their inhabitants a common consciousness and local civic spirit. Nor have exclusive workplaces, as a rule, proved of value as foci of community spirit. Certain notable exceptions from this rule will be dealt with in a subsequent paragraph.

3 *INTERPRETATION OF THE SOCIAL FAILURE, SO FAR, OF THE SEVERANCE*

The disturbing discrepancy between the potential and the actual development calls for an explanation. Does the unfavourable diagnosis of the present state of affairs inevitably involve a pessimistic prognosis? Fortunately, this is not the case. The experiment in urbanism of severing the dormitories from the workplaces need not be considered to be merely disastrous. A survey of the reasons for the comparative failure of dormitories and exclusive workplaces indicates that to a certain degree the faults are capable of remedy, whilst other mistakes which have been made may be avoided in future development.

a Newness of the Dormitories

Turning first to the dormitories, it must be kept in mind that none of the schemes built since 1919 had been in existence for twenty years when the present war broke out. A few decades, however, are a very short time in the life of an urban settlement. Considerable periods of infancy must be reckoned with, before towns attain the stature of mature urban communities in the full sense of the word.[1] The dormitories, and even some pre-1914 suburbs which have not existed for more than one generation, have not yet had the chance to prove their mettle.

This does not mean, however, that the special needs of the new dormitories have been fully realised and provided for from the start. In places of slow organic growth, community feeling and civic spirit may perhaps be expected to develop naturally with the growth of the place, although even there a helping hand, in furthering clubs and the like, seems desirable. But the rapidity with which big modern dormitories spring into being calls for the making of deliberate efforts to inspire a common consciousness immediately upon the opening of the new housing

[1] See for instance, J. R. Maud, *City Government. The Johannesburg Experiment*, 1938 passim.

estate. Mr. Terence Young has shown for Becontree,[1] and it is now generally realised, that the new dormitories have been starved of community life during the initial years. With the notable exception of the churches, there was little opportunity for, and stimulation of, social and civic life before community centres were started. Community centres assist and supplement clubs and similar organisations with the aim of developing a better and fuller life for the community and for its individual members. It is a difficult task now to introduce common consciousness into the existing housing estates after the opportunities of the early formative years have been missed and indifference towards the community allowed to take root. In future, it can be counted upon, the mistake will not be repeated.

b Thinness of the Urban Fabric

Even so, the further development of the dormitory suburbs cannot be awaited confidently without reform. One particular feature of their arrangement is a serious obstacle on the road to social integration, and that is one which cannot easily be altered in existing areas. Open development of the suburbs, which has been the rule since 1918, must be held partly responsible for the isolation in which many suburban families live. Where houses are laid out at twelve to the acre, neighbourliness does not thrive as easily as in more compactly built districts, and it is more difficult to provide statutory and voluntary community services.[2]

c Novelty of the Severance: Maladjustment of Organisation

Yet newness and thinness of the urban fabric are only part of the explanation. Dormitory suburbs, not so new and not laid out at low density, have nevertheless failed to attain the full life of an urban community, as is shown, for example, by the Merseyside borough of Wallasey. There is a third point which is particularly important, in view of the fact that here reforms are both necessary and possible. The dormitories are not only

[1] *Becontree and Dagenham*, 1934 passim.

[2] On the strength of these arguments, a plea for maintaining the distinction between town and country has been made by Mr. W. Harding Thompson. Reported in *Town and Country Planning*, Vol. 1940, p. 32. Other implications of a low density of building will be discussed in detail in the subsequent chapter on town planning.

young as urban units, but represent also a novel type of urban settlement. The substitution of two specialised localities for one comprehensive community is an innovation which calls for careful adaptation of the administrative and social framework. Actually, however, little attempt has yet been made to find the proper organisation of urban life under the conditions of severance. This omission seems to be a major cause of the non-achievement of social integration in the new areas. The following discussion can only be tentative, since the development is as yet in its initial stages.

The dualism of places plainly corresponds to the two primary human concerns: the dwelling place is the stage of private and family life; to the workplace, on the other hand, man is attached as the source of his livelihood and the focus of his vocational interests. Besides home and work, however, there exists the whole sphere of community life: statutory and voluntary services and a variety of leisure-time pursuits which have become important components of modern society. They form the objects of "associations of many different kinds . . . (whose) multiplication is one of the features of a modern civilised community".[1] Where is *their* place in the new urban arrangement? To this question the present organisation of community life does not allow a satisfactory answer. It would be possible to allocate the activities of community services and associations to a third place, coincident with neither dwelling place nor workplace. Rudiments of such development are to be seen in the "Theatre Land" of London, in the concentration of cinemas in the centres of some medium-sized towns and in the nucleated arrangement of the buildings of some newer universities.[2] The future may bring a strengthening of such "character zoning", but the urban lay-out as it is at present shows in the main a different design. Social machinery has been set up at the places of residence and of work respectively. Its allocation, however, does not reflect a realisation of the importance of proper assignment, either by public authorities or by voluntary associations. On the one hand, local government status is being withheld from the new areas, and on the other hand, nearly all services that

[1] A. M. Carr-Saunders and D. Caradog Jones, *Social Structure of England and Wales*, 1927, p. 83.

[2] To these instances may be added the shopping centre, which is in many cases situated at a distance from both dormitory and workplace.

have been set up are situated in the dormitories at the expense of the workplaces.

i WITHHOLDING OF LOCAL GOVERNMENT STATUS

"At no point in the modern history of London has the machinery of local government been adjusted so as to take the facts of population into account."[1] This statement is true also of other towns. Older suburbs were only belatedly given the status of boroughs, and the housing estates built after the last war have not, so far, been made areas of local government; they are "left without separate legal existence or organisation, resulting in a lack of cohesion and a diffusion of responsibility".[2] Most of the housing estates, however distant from the city's town hall and however numerous their population, are "simply a physical part of the city",[3] and the L.C.C. estate, Becontree and Dagenham, as has often been pointed out, belongs to three different non-London local authorities. There can be little doubt that the social spirit of the dormitories suffers from the fact that they do not form local-government units, from the lack of "common government (which) is itself a powerful unifying force setting a common stamp upon an area and its inhabitants".[4] This is demonstrated by the development of Wythenshawe, which is within the city boundaries of Manchester. As its chief promoter complains, "there is no community life growing up . . . it is simply a suburb of Manchester".[5]

The endeavour to make up for this deficiency has found expression in the movement for community centres. A community association (which backs the community centre) is described by Dr. Barker as "a sort of half-way house between the unit of local government and the ordinary type of voluntary association".[6] Community centres have an important function to fulfil, but it can hardly be expected that the secretaryship of a dramatic society or other club will call forth the same sense of

[1] W. A. Robson, *The Government and Misgovernment of London*, 1939, p. 49.
[2] Sir Wyndham Deedes, Chairman's address at a conference of the National Council of Social Service, etc., 1938, Roneographed Report, pp. 18–19.
[3] Dr. Ernest Barker, Preface to *New Housing Estates and their Social Problems*, published by the National Council of Social Service, 1935, p. 3.
[4] H. A. Mess, loc. cit., p. 397.
[5] Sir E. Simon, Evidence before the Barlow Commission. Paper No. 73A (unpublished), p. 25.
[6] Loc. cit., p. 4.

responsible citizenship as is engendered by serving on a Local Government Committee.[1]

It follows that the dormitories should be made units of local government as early as possible.[2] More than that, the local authorities of these new, developing areas require different and in some respects extended powers of self-government, as compared with older districts.

Severance is accentuated where dormitory and workplace are not incorporated in the same larger municipal unit. A wider local government area comprising the correlated places tends to attract some of the local patriotism and civic consciousness which were formerly devoted to the self-sufficient town. The severance of home and workplace is thus a point in the case for regional organisation. But although regionalism may render valuable help in cultivating local tradition, it cannot attain this aim by itself. Development of local community life remains of the utmost importance.

ii UNDISCERNING ALLOCATION OF THE SOCIAL MACHINERY

In the distribution of the social machinery, dormitories show a remarkable predominance over workplaces. It is difficult to evade the impression that in the setting up of the various services, the problems raised by the severance of the places have been disregarded : the traditional procedure has been indiscriminately maintained, namely of establishing the organisations where the people live.

The problem of proper location affects the various organisations to a different extent. There can be no doubt, for instance, that elementary schools have to be near the children's homes, nor that membership of consumers' co-operative societies rightly continues to be based on residential neighbourhoods.

For the trade unions, on the other hand, the question whether to locate their branches at dwelling places or workplaces constitutes an administrative problem of the first order. The implications of the severance of home and workplace are making themselves felt. The "obstacles in the way of the organiser are aggravated . . . by the fact that the homes of the workers

[1] A comprehensive description of community centres was published in 1942 : F. and G. Stephenson, *Community Centres*, a Survey for the Community Centres Joint Research Committee.

[2] See H. A. Mess, loc. cit., p. 406.

are frequently scattered over areas far removed from their place of employment. . . . It is very difficult for a factory organiser to meet men outside working hours . . ." [1] The traditional organisation of trade-union branches still prevailing amongst the greater part of British industry is the grouping on the basis of residence. This principle of organisation was adopted at a time when workers lived in vicinity of their work and therefore also in vicinity of their work-mates, and when working-class districts covered comparatively small areas. To-day, circumstances being different in all these respects, trade-unions organised on the basis of residence find it frequently difficult to ensure an active participation of their members in the union's life. Nor is a solution simply to be found by holding the meetings near the works or offices, following the example of the printers' chapels, for workers living at a distance often fail to turn up at evening meetings held near the factory. One solution might be to have the meetings near the workplace following on after working hours. Another solution would be to meet in the evening at a civic centre within easy reach of various residential areas.

The question where the local meetings ought to be held requires reconsideration also in view of another development. Originally, the trade union concerned itself with both working and private life of its members. The transition of most of the friendly society functions from the unions to the State, however, has lessened the trade union's concern with the member's private circumstances and shifted the emphasis even more than before to those activities that are connected with the workplace.

The matter is complicated by the fact that behind all questions of trade-union reorganisation there looms large the fundamental issue whether the grouping should be based on craft or on industry or on the workshop. This problem is outside the compass of the present study. The severance of dwelling place and workplace is only one of various factors to be accounted for in developing the framework of trade unionism ; but the question where local meetings are to be held is important for the vigour of the movement and one that deserves more attention than it seems to have received hitherto.

Imperfect cognisance of the implications of severance is also

[1] John Parker, " Trade Union Difficulties in New Areas " in G. D. H. Cole, *British Trade Unionism To-dav*, 1939, p. 241.

visible in the organisation of the Employment Exchanges: the unemployed are expected to register at the Local Office nearest or most convenient to their places of residence, while the Local Employment Committees are meant to bring the Exchanges "into closer contact with the local industrial community".[1] But there is no "local industrial community" if, for instance, a man lives in Morden or Watling and used to work in Battersea or Slough. A more adequate organisation may be expected to result from further experience and research.

At present, only few activities in the sphere of community life are to be found at the workplace. Adult education is one of them. The Workers' Educational Association holds classes at both workplace and dormitory, in order to suit the different needs and interests of the students in their respective capacities as workers and as residents (the latter include non-earning housewives as well as earners). At Bournville, the juveniles employed by Cadbury Bros. attend a day continuation school one day each week instead of working in the factory.

Recreational activities at the workplace are another feature of social life which has gained increased importance during recent years. It has become more and more the habit of firms to provide for sport and other leisure-time pursuits of their employees near the factories. It is not known how many firms organise recreational facilities, but "the number of schemes must be very large and the numbers using them considerable".[2] These schemes, it is maintained, have grown up in response to an obvious need: "There is nothing artificial in firms undertaking the organisation of such activities, for in the majority of cases they have done so in the absence of others being available and in response to sustained demands on the part of their employees" (p. 7). The community-building influence of these clubs often extends beyond the firm's employees, for "in an increasing number of schemes, members are allowed to bring their family and friends . . . a development in every way to be welcomed" (p. 9). Such activities obviously foster comradeship, community feeling and allegiance; but in present circumstances, it is mostly single large undertakings and not groups of firms that

[1] *Ministry of Labour Report for 1938*, pp. 18–19.

[2] *Recreation in Industry*, published by the Industrial Welfare Society Inc., 1938, p. 6.

cater for their employees' leisure-time activities. This fact makes it difficult to say to what extent the common consciousness and loyalty accrue to the place of employment, as distinct from the employing firm. It will, therefore, be interesting to observe the development of Slough Social Centre, which is organised on an unusual basis. The Trading Estate Company and local manufacturers have co-operated in the establishment of this scheme of varied recreational facilities.[1] The Centre is open to all Trading Estate employees and their families living within a ten-mile radius and is, therefore, a feature of the workplace and not of a particular firm.

However, residents of the old town of Slough who are not employed on the Estate can equally become members. Some of them and of the inhabitants of the nearby housing estate have done so. The Slough Social Centre contains, therefore, elements of both a residential and a workplace grouping.

Exclusive workplaces like Longbridge (founded in 1905) or Cowley (1912) show no tendency to develop a specific local spirit, although the works have been on the same sites for several decades. Local community sense is equally wanting at Trafford Park Trading Estate, which was founded over forty years ago; and Slough Social Centre, it is well known, owes its creation to the effort to amend unsatisfactory industrial relations and to reduce the high labour turnover hitherto prevailing at the Trading Estate.

Yet there are other exclusive workplaces which display a vigorous common consciousness. The City of London is the main example, but other pure business districts, for instance that centred around the Cotton Exchange and Stock Exchange in Liverpool, produce on a minor scale the same phenomenon as that found in London. How is it to be explained that community spirit prevails in some exclusive workplaces and is wanting in others? There seem to be two important factors: firstly, the City of London and similar workplaces have grown out of old compound communities, inheriting and carrying on a strong tradition of citizenship. The other group of workplaces, such as Trafford Park and Longbridge, were started without antecedents of that kind. The argument previously put forward

[1] Various public authorities contribute towards the running costs. See *The Times*, 29th September, 1937, and Barlow *Report*, p. 284.

concerning the newness of dormitories applies to a certain extent also to factory districts and trading estates. Secondly, the two groups of workplaces differ in the types of vocations and activities carried on. Exclusive workplaces attain a community spirit when they are peopled by persons with professional interests, or what may be called such. Where a market or financial centre exists, there is something like a professional atmosphere, resulting in the development of common consciousness, as shown outwardly by gathering in particular eating places (as distinct from works canteens), clubs, and so on. Where those who gather in exclusive workplaces are wage-earners only, no such spirit arises.

As a rule, the two features occur together, that is to say, markets and quasi-professional activities are to be found in old-established places with a tradition behind them, and not in new workplaces. Local community life does not, however, invariably attend or follow the transformation of a mixed neighbourhood into a mere business district. The shopping centre in the West End of London, which has developed out of a mixed district but has no market or exchange, displays no local spirit comparable to that of the City. This shows the importance of the " professional atmosphere ". The question is whether social integration of the workplace can be promoted only by persons with professional interests.[1] Cannot similar services be rendered to industrial workplaces by workers' associations and other activities? It seems well worth while to strive for this end, for " desire to associate and ability to do so are the prerequisites . . . not only of professionalism but also of any society which is vital and free ".[2]

While the poorness of community life in the dormitories has recently met with a great deal of anxious attention, the same state of affairs in exclusive industrial workplaces is largely accepted as natural and unavoidable. The above-mentioned educational and recreational activities near the factories show, however, that improvement is possible. The factory districts lack of tradition and the instability implied in labour turnover are serious handi-

[1] H. Durant relates such a favourable state of affairs to the satisfaction many professional men, artists, merchants, and skilled workers obtain from their work. " Finding satisfaction in work, they do not desire to flee from it as soon as the immediate job is completed . . . There is for them no sharp break . . . they have tended to build up leisure time associations on a vocational basis ". *The Problem of Leisure*, 1938, p. 258.

[2] A. M. Carr-Saunders and P. A. Wilson, *The Professions*, 1933, p. 495.

caps ; but not all of the factors which at present obstruct the cultivation of local spirit at the industrial workplace seem inevitable. Problems of the workplace are largely identical with problems of industrial organisation and industrial relations. Consideration must, therefore, be given to such factors as the discouragement or even prohibition of trade-union activities near the workplace by some firms, and workers' mistrust of activities undertaken by the employers ; another relevant factor is the attitude of public authorities towards the civic development of the workplace.

The question whether the elements of social and civic life at the workplaces can be expanded and strengthened is very important. The success of the severance of dwelling place and workplace depends on it. Severance cannot produce a specialisation in the ways of living, and thereby a more efficient citizenship, if in one of the two localities concerned community life is all but eclipsed. In that case, the inhabitants of the dormitory would be relegated to a fragmentary citizenship by being debarred from participation in the potential interests of the workplace. This would mean a weakening instead of a differentiation of citizenship. The social damage caused thereby would, in the long run, be infinitely greater than the direct costs and hardships of the daily journey.

4 *CONCLUSION*

" The possibility of severing workplace from dwelling place is a very important social fact." [1] The use made of this possibility not only influences the ways of living of individuals but changes also the face of urban settlement. Originally, the severance of the two places was of a rather incidental nature, occasioned by journeys to work which were the outcome of conditions of the labour market and of family life. This state of affairs still exists to a large extent : the inhabitants of older villages and urban districts may or may not find employment near their homes ; single persons and families may or may not move nearer to the workplace—whoever works outside his domicile undergoes the personal experience of a " divorce be-

[1] A. C. Pigou, " The Concentration of Population ", in *Essays in Practical Economics*, 1924 ed., p. 103.

tween residence and workplace ". The position is different with outlying housing estates on the one hand and business centres and factory districts on the other. These forms of lay-out entail the severance of places as a matter of course, partly even as a contemplated design. In the shape of dormitories and exclusive workplaces, two specialised localities take the place of the former comprehensive community.

The tendency towards such a polarity of the urban structure is bound to affect the quality of community life. This may be to the good or it may not. Specialisation, in this as in other fields, is capable of generating a salutary differentiation and thus a fuller social and civic life ; but specialisation may also lead to social disintegration and civic indifference. Unfortunately, the latter course has, so far, been prevalent. What are the causes of this undesirable development, and what conclusions are to be drawn for future policy ? The main reasons for the social shortcomings of severance are : the newness of the dormitories, the thinness of the urban fabric, brought about by open development of the suburbs, and maladjustment of the administrative and social framework to the novelty of the severance. The potential beneficial effects of severance have thus been frustrated by mistakes which are not altogether unavoidable.

Reforms have to start at various points. The most conspicuous need is for the development of common sentiment and civic spirit in the dormitories. The somewhat sporadic efforts made towards amending the "atomistic condition" of the residents have to be sustained, intensified and expanded. The provision of communal leisure facilities and decentralised services should be supplemented by making the residential districts areas of local government. Town-planning policy, too, will have to take into account the experiences of the last two decades. Another important requirement is to foster citizenship at the workplaces. Community life in factory districts is, as a rule, very weak. But society cannot afford to neglect the workplaces as potential foci of common interest and local affinity. A considerable part of the earner's time and energy is spent at the workplace. This being so, his community interests, too, should be divided between dormitory and workplace. An expansion of vocational and recreational activities at the workplace would help towards creating a common spirit at industrial workplaces.

similar to the atmosphere which prevails in some commercial and financial centres.

Measures of these various kinds have to be applied on a considerable scale and over a number of years before the results can make themselves felt. If accompanied by a proper organisation of social and civic life, the severance of dwelling place and workplace may well prove to be for the benefit of the urban population.

Chapter V

THE DAILY JOURNEY AND COMMUNITY LIFE

II Implications for Town Planning

THE journey to work is a test of the urban lay-out : excessive costs and hardships of daily travelling can be taken as a symptom that something is wrong with the form of the town or conurbation. Location of industry plays, of course, an important part in this respect, but, as mentioned before, that is a topic beyond the scope of our investigation.[1] We are here concerned with the size and pattern of the town with regard to the homes of the people. The two factors, size and pattern, are interdependent. This fact can be seen to underlie various different suggestions as to the optimum size of towns : Ebenezer Howard, in adopting a very low density of building, limited the number of the town's inhabitants to 30,000. Le Corbusier, on the other hand, arrives at his figure of 3,000,000 by way of designing blocks of flats of skyscraper type. Policy and public discussion of town planning, however, have lost sight of the importance of the interdependence, with the result that the issues have become obscured. It seems, therefore, useful to examine these two factors of the form of towns in the light of the journey to work.

1 *THE SIZE OF THE TOWN*

The analysis of the functions of the journey to work showed the merits of a large town from the point of view of the labour market, in the interests of both economic and family life. The chapter on the price paid for the daily journey demonstrated the drawbacks. A compromise must, therefore, be sought between the facility of choosing jobs and employees, on the one hand, and the waste of lengthy journeys, on the other hand. Up to the present, however, the growth of the towns has gone unchecked.

[1] See, however, pp. 14/15 and p. 108 ff.

The term " size of town " has a twofold meaning : it refers to numbers of inhabitants and also to square miles of built-up area. In fact, the growth of the towns, once the old boundaries were overstepped, has proceeded in both ways. The " world phenomenon of urbanisation " [1] has taken the form of aggregations of millions of people in large units, as well as of a " spread (of the cities), overflowing their boundaries and forming sprawling agglomerations of humanity ".[2]

From the point of view of the journey to work in particular, but also of the lay-out of the town in general, the spatial expansion is of signal importance. This has not, however, been duly appreciated in the literature and practice of town planning in recent decades. " Size of town " is commonly applied to the number of inhabitants only, and the extension of the built-up area is regarded as a secondary feature, as a mere by-product of the increase in population. While the latter is given attention with a view to regulating it according to the needs of society, the expansion in space, however much lamented as hypertrophic, is usually considered as not capable of being checked by itself. On the contrary, " London is spreading even more quickly than it is increasing ",[3] and this also holds good of Birmingham and other large towns. Suburban spread has been the deliberate policy in the last twenty years, and has been welcomed as a measure to relieve the excessive density of population in the old districts of the towns. There is thus created an antagonism between the just demand of housing reformers for the loosening of the urban fabric and the equally justified complaints about the ever-growing urban expansion. The size of the built-up area of the conurbations is objectionable for reasons from without as well as from within ; the bewildering unwieldiness of the large town itself finds its counterpart in " the alarming way in which the countryside is being overwhelmed ".[4]

No figures are available for a comparison of the growth in numbers of population and the growth of the acreage " covered by bricks and mortar " at the various stages of urban growth.

[1] Royal Commission on the Distribution of the Industrial Population, *Report* (1940, Cmd. 6153), in the following quoted as Barlow *Report*, p. 10.

[2] Ibid., p. 11.

[3] Association of Road Operators Ltd., Barlow Commission, *Evidence*, p. 685.

[4] *The Land of Britain.* The Report of the Land Utilisation Survey of Britain, Part 79. Middlesex and the London Region, by E. C. Willatts, p. 168.

" Greater London " has since its first use meant the Metropolitan Police District, that is practically the same area of just under 450,000 acres which it covers to-day; the built-up area, as distinct from the total acreage, has never been measured. The general unconcern prevailing until quite recently about urban spatial expansion is reflected in the ratios between " Population and Acreage, 1801–1931," in Greater London given in the Report on the Census of England and Wales, 1931.[1] The population figure of each Census year is related—not to the built-up area in that year, but to the total acreage of Greater London as it was in 1931. The ratio of " persons per acre " is thus automatically shown to rise in the same proportion as the number of inhabitants, multiplying by over sevenfold between 1801 and 1931.

Some attempts have been made to map the growth of London's built-up area during the last century.[2] The charts give an impressive picture of the increasing spread of Greater London, but in the absence of figures it is not possible to compare the spatial growth with the numerical growth of the conurbation.

For " The Birmingham and Black Country Area ", on the other hand, the growth of the built-up area in the last twenty-five years has been computed, but the figures are unaccompanied by data on the corresponding growth of the population. The built-up area in this region has, from 1913 to 1938, increased by 68%. For Wolverhampton and Environs alone, the increase is over 71%, for Coventry and Environs, over 83%.[3] The population of the City of Birmingham increased in the same period by less than 25%.

The changed proportion between the number of inhabitants and the built-up area has been of great influence on the shape and functioning of the towns. This question will be examined in detail in the subsequent section on the density of building. Without consideration of this important point, all discussions on the optimum size of the town are lacking in substance.[4]

[1] *County of London*, Table 1.

[2] L.P.T.B., *Third Annual Report*, front and back covers.—Steen Eiler Rasmussen, *London the Unique City*, 1937, p. 134 seq., reproducing another set of diagrams supplied by the L.P.T.B.—*London Statistics*, 1936–8, map.

[3] *When We Build Again*, p. 132 seq.

[4] See, for instance, p. 102, note 3.

2 *THE PATTERN OF THE TOWN*

a The System of Lay-out

i peripheral growth

Even if the spatial expansion had been only parallel to the increase of the urban population, the continuous growth of the large towns would have led to serious problems, as, for instance, the difficulty of access to the countryside for the inhabitants of the central districts. Yet the Barlow Commission pronounces that " the natural place for a town to provide additional houses to meet the needs of its population is on the outskirts of its built-up area ",[1] and, in fact, peripheral expansion was for a long time the only form of urban growth in this country. New suburbs grew in immediate, or nearly immediate, contiguity with the old districts. The suburbs were largely of a residential type ; only in the last two decades have many factories, too, moved out to the fringes of the towns. The development of peripheral expansion has gone to such lengths that the large towns, quite apart from the danger of choking the centre, are becoming unmanageable.

The haphazard and unsatisfactory manner in which London and other conurbations have grown is now generally recognised. Nor is it necessary here to expatiate on such features as the absorption by the central town of hitherto separate townships which ultimately coalesce,[2] or the moving-out of people to live " 'beyond the suburbs', thereby creating fresh suburbs even further out ".[3] However, a review of the town-planning problems from the point of view of the journey to work can contribute to an appreciation of the current suggestions for planning the towns of the future.

Ought the former policy of *laisser-faire* to be replaced by the other extreme, coercion ? It seems a counsel of despair to propose the restriction of the freedom of migration or, even more sweeping, the removal of a large number of the inhabitants from London and other large cities and their settlement in entirely new towns. Nor will a policy of decentralisation of population and industry

[1] Barlow *Report*, p. 68.
[2] Greater London Regional Committee, *Second Interim Report*, p. 93.
[3] E. C. Willatts, loc. cit., p. 166.

suffice to solve the problems of London and the other conurbations. Even if carried far, decentralisation is not likely, for many years to come, to cause a melting away of the large towns.[1] The need is for constructive planning of the area occupied by the large town and its surroundings.

There is now a large measure of agreement that the growth of towns should not proceed by continuous peripheral expansion but should be founded on the " unit system of planning ". The most elaborate conception of such units is that of the satellite town.

ii SATELLITE TOWNS VERSUS GARDEN CITIES

In the discussion of town and country planning, the term " Satellite Town " has tended to become confused with the older one of " Garden City ". This is apparent from the following statement of the Greater London Regional Planning Committee.[2] " The satellite town or garden city seems to offer the most complete and effective ultimate type towards which to work." The Marley Committee even pronounced explicitly that it was " questionable whether at this date there is any great value in the maintenance of the expressions in any definite sense ".[3] The Barlow Commission does not go as far as that ; its Report (p. 137) implies that sometimes either garden cities or satellite towns should be encouraged ; but there is no distinction made between the two terms. The difficulties in the current discussion on these matters are partly occasioned by the fact that the terminology has not kept pace with the actual development of town planning.

A garden city in the original sense is a new independent centre laid out at a low density of building. In the course of time, however, these principles have become disconnected. On the one hand, " suburban development on garden-city lines " or " the garden-city principle applied to suburbs " do not create independent towns. The term " garden city " has thus been watered down through the inclusion of " garden suburb ". On the other hand, the project of a new independent town is not necessarily a project of development at a low density ; the scheme of the Hundred New Towns Association, for example, provides

[1] See Sir G. Gibbon, *Reconstruction and Town and Country Planning*, 1942, p. 96 and passim.

[2] Loc cit., p. 109.

[3] Departmental Committee on Garden Cities and Satellite Towns, *Report*, 1935, p. 6.

for compactness of building.[1] Both types have sometimes been called "garden city".

A satellite town (of which no complete example exists) is another idea of new planned town. As the name indicates, it would to some extent be dependent on a large town but at the same time have a social, civic and economic life of its own. A belt of open country would separate satellite towns from the central city and from each other. Such an urban arrangement has been described as sub-centralisation.[2]

While Letchworth and Welwyn Garden City combine elements of the garden city and of the satellite town, general urban development during the last two decades has proceeded almost entirely on garden-city lines and ignored the satellite-town idea. The opposite course would have been the right one, in the interest of the journey to work and indeed of proper planning in general.

The satellite town helps towards mastering the problem of large agglomerations, while the garden-city policy of open development adds to the difficulties. The building of satellites is contrasted with a continuous peripheral expansion or "sprawl" of a town. Satellite Town is a term of integration; it refers to the grouping into an organic system of otherwise amorphous masses of buildings. Garden City is a term of distribution. It refers to the provision of plenty of open space within the built-up area for private gardens and wide promenades and to the decentralisation of the industrial population. But decentralisation is up against a powerful tendency: "The call of the . . . big city . . . represents a tide of forces, social as well as industrial in character . . . likely to be difficult either to dam or to direct . . ."[3] To the neglect of this tendency and to the looseness of the urban fabric must largely be attributed the moderate success of the garden cities in building up full urban communities; the ties uniting the inhabitants of such places consist largely in conceptions of life held in common before they moved to the garden city rather than in the integrating force of the locality itself. In the words of the Barlow Commission,[4] Letchworth (1903) and Welwyn (1920) were "promoted . . . as experiments in social living. Their appeal was largely to those in sympathy with the ideas of the founders; this some-

[1] *One Hundred New Towns*, Anonymous, 1934.
[2] Thomas Sharp, *Town and Countryside*, 1932, p. 172.
[3] Barlow *Report*, p. 13.
[4] Ibid, p. 128.

what limited appeal . . . may account to some extent for the comparatively slow growth of these two communities."

iii EXPEDIENCE OF SATELLITE TOWNS

From the point of view of the journey to work, the development of satellite towns, that is, of circumscribed urban units near a large parent town, is definitely beneficial, and this for two reasons; firstly, fewer persons have to undertake daily journeys, and secondly, the journey for those who have to travel to their work, either in the central town or in another satellite, is rendered both cheaper and quicker.

A satellite is, by definition, to a certain extent self-contained, that is to say, it includes in its comparatively small unit a variety of workplaces as well as residential quarters. Consequently, the majority of earners can work within walking or cycling distance of their homes; the total of daily journeys is thus reduced by the number of persons who find work locally. These earners have their lives again centred at one place which provides them with both home and employment, instead of double-focused at the places of residence and of work. In so far as the severance between dormitory and workplace is found to impede the development of citizenship and community spirit, the persons "localised" in a satellite town may, therefore, serve as a nucleus for the growth of civic patriotism.

At the same time, easy access to the large parent town and to neighbouring satellites satisfies the demand for that mobility of labour, through daily travelling, which was found necessary for social and economic reasons. It further enables the inhabitants to keep contact with, and to participate in, the various cultural opportunities which only a large town can provide.

Within a given area, a grouping of urban clusters, leaving free large green belts, can be served by a railway line with fewer stations than would be needed for the same number of people scattered over the whole area of the conurbation. The operation of stations, however, with their costly equipment and staff, is a considerable item in the expenses of a railway.[1] The running costs, too, are reduced by a reduction of stops such as is made possible by a lay-out in circumscribed urban units, as contrasted with a thin spread of houses over the whole area. "There are

[1] See p. 101.

certain economies in running non-stopping trains, since maximum speed need not be so high and the cost of stopping such trains, which varies from 3*s*. to 6*s*. a time, is eliminated."[1] Another aspect of the same argument holds good, if the emphasis is laid on time instead of on cost: train services of equal maximum speed are faster with fewer stops, for "it takes a mile or so to attain any great speed".[2] The situation may be summed up in Professor Abercrombie's words: "Mere proximity does not make for quickness of access, as . . . there is a tendency for frequent stops; in London to-day there are many places outside the continuous growth which by means of an express service are within a shorter time-distance of the centre."[3]

b Density of Building

i THE ASCENDANCY OF OPEN DEVELOPMENT

The low density of twelve houses to the acre, which has become the standard for the huge housing output since the last war, is transforming the face of urban settlement: the former compactness of building has given way to open development of vast surburban areas. Three factors have combined to bring about this sweeping change: the motives had existed for years, viz. a changed conception of "proper sanitary conditions, amenity and convenience";[4] the means were steadily improving, by the development of modern transport services; the opportunity arose when house-building for the masses of the population became dependent upon public subsidies and thus subject to direct instructions from the Government.

The Motives.—The case for low density of building is founded on considerations of social progress as well as of public health. The foremost argument, however, is that of health. The deplorable living conditions in the extensive slums of the big towns were the concern of housing reformers throughout the greater part of the nineteenth century. The main features of the slum are congestion and lack of sanitation. Towards the end of the century, much leeway had been made up in regard to sanitation;

[1] K. G. Fenelon, *Railway Economics* (1932), p. 164 (the actual figures refer to long-distance express trains, not to suburban trains).

[2] F. J. C. Pole and James Milne in *Modern Railway Administration* (1925), p. 90.

[3] *Town and Country Planning*, First Edition, 1933, p. 119.

[4] *Housing and Town Planning Act*, 1909, Section 54(1).

but the congestion remained grievous—in the living quarters in the central districts, the penetration of air, light and sun was quite insufficient ; open spaces all but non-existent ; and overcrowding rampant. The obvious remedy was loosening of the urban fabric.

Together with health objectives, the promoters of open development have "social betterment" in view. From the beginning of this century, town planning on the basis of low density has been demanded as a measure against "the body-blighting, soul-destroying conditions of overcrowding and the slums".[1] And action was promised "not only in the interest of public health and of the comfort of the general community, but also of their moral and spiritual welfare".[2] In the name of social equality, each family was to be given a cottage with a garden of its own. This former privilege of the well-to-do was now thought to be within reach of the masses of the population. "The garden suburb of the wealthy was at least a century old. It was now time to democratise it."[3] The same conception of the town-worker's cottage as a separate social and architectural unit led to the detachment, or at least semi-detachment, of the houses and to the setting back of the buildings from the street behind a deep building line.

Beauty has been another object of the advocates of low building density. "Instead of gloomy streets and squalid dwellings", instead of "the depressing effect of monotonous unbroken rows" the aim was to "develop beauty of vista, arrangement and proportion."[4] In the words of Professor Abercrombie, "one of the chief advantages of the lower density in modern suburban planning has been the possibility of introducing beauty, which here stands for the preservation of trees and greenery, an improved type of domestic architecture, the avoiding of monotony, and the planning the whole site to group houses together".[5]

Ways and Means.—From the beginning of the present century, the technical means for a widespread lay-out of the conurbations have developed spectacularly, through the unprecedented growth of modern transport. The network of local transport lines—

[1] H. R. Aldridge, *The Case for Town Planning*, 1915, p. 166.

[2] Reply of the Prime Minister, Sir Henry Campbell-Bannerman, to a deputation of the National Housing Reform Council, 1906. H. R. Aldridge, loc. cit., p. 182.

[3] P. Abercrombie, loc. cit., p. 86.

[4] *Tudor Walters Report* (Cd. 9191, 1918), paragraphs 53 and 56.

[5] Loc. cit., p. 106.

trams, electric railways, buses and trolley-buses—was continually extended and more closely knit, bringing ever wider areas within daily reach of the town centres. Technically, indeed, transport can cope with the suburban spread involved in open development.

The question of the economic ways and means was less simple. To the promoters of low density, the problem presented itself as an increase, in two ways, of the costs of building : the price of the greater acreage of land needed for a given number of houses, and the costs of developing the larger site. How could the community afford open development in the face of these financial difficulties ?

The first (Housing and) Town Planning Act, 1909, had marked a break-away from the policy of the *laisser-faire* period which had allowed the utmost use of building land, almost without regard to health and amenity. The Act, within its restricted scope, provided that " property shall not be deemed to be injuriously affected by . . . provisions in a town planning scheme which . . . limit the number of buildings to be erected ".[1] Limitation of the number of houses to the acre involved, therefore, " a nice adjustment between the claims of the landlord for a fair return on his capital and the claims of the community for means of a healthy and decent house to live in within its means ".[2] As to the owners of potential building sites, it was argued that " landowners can afford to sell land for low density more cheaply than for high, because they sell more acres for every thousand houses ".[3] As to the interests of the community, we need not here enter upon the intricate problems of land-values, compensation and betterment ; suffice it to say that, relying on the services of transport, the purchasers of building land got ever farther out from the town centres and thus secured sites at reasonable prices.

For dealing with the second point, the cost of developing suburban areas at low density, the Local Government Board appointed in 1917 a committee " to consider questions of building construction in connection with the provision of dwellings for the working classes and report upon methods of securing economy and despatch in the provision of such dwellings ". The Report

[1] Clause 59, sub-section 2.

[2] George Cadbury Jr., *Town Planning*, 1915 ed., p. 111.

[3] Sir R. Unwin, *Nothing gained by Overcrowding*, Garden City and Town Planning Association, Leaflet 13, 1932. See also note 1 on p. 101.

of this Committee (known as the Tudor Walters Report [1]) can be regarded as the charter of open development. It came to the conclusion that " no addition to the cost of development need generally be involved on account of such limitation " to twelve houses per acre, if only " the lay-out plan is carefully prepared ".[2] In order to achieve this result, the plans recommended by the Committee replace streets by narrow carriageways and footpaths.

In this manner, the ways and means, both technical and economic, were deemed adequate to the ends of urban expansion. Consequently, the Tudor Walters Report envisaged a " bold and enlightened policy " with the aim of creating " spacious suburbs ", where " the population is spread over a large area and ample gardens and other grounds are provided ". The Report further laid down that " it is most desirable that a rather more generous frontage should be allowed . . . than has been adopted generally in the past. Slight additional expenditure should not be regarded as justification for restricted building frontage ".[3]

The more fundamental implications of low density of building, other than financial, have found but occasional mention in town-planning literature. They will be discussed at length in the next section.

The Opportunity.—Two decades before the opportunity arose to carry out open development on a large scale, there was a pioneer movement blazing the trail. It is the historical merit of the Garden-City movement started by Ebenezer Howard,[4] to have challenged the congestion and ugliness of the old town by a new conception of urban settlement.

Low density of building is a matter of course in the garden city. In the formal definition of Garden City, adopted at a time when open development had become the accepted policy of the Government, building density is not even mentioned.[5] Letchworth, the first garden city (founded in 1903), is built at a density of only six houses to the acre ; [6] it had in 1939 a population of

[1] Cd. 9191, 1918. [2] Ibid., para. 351, subsection 7. [3] Ibid., para. 74.

[4] *To-morrow*, 1898 ; renamed in the 1902 edition *Garden Cities of To-morrow*.

[5] The Garden Cities and Town Planning Association adopted, in 1919, the following definition : " A garden city is a town planned for industry and healthy living, of a size that makes possible a full measure of social life but not larger ; surrounded by a permanent belt of rural land, the whole of the land being in public ownership or held in trust for the community." C. B. Purdom, *Town Theory and Practice*, 1921, p. 34.—The Association recently changed its name into " Town and Country Planning Association ".

[6] Sir R. Unwin, Barlow Commission, *Evidence*, p. 846.

17,000. Welwyn Garden City, founded in 1920 and equally promoted under private auspices, numbered about 15,000. Two other garden cities, Whythenshawe and Speke, have been developed by the Corporations of Manchester and Liverpool respectively. Before the present war, the former had 35,000, the latter, still in an initial stage, a few thousand inhabitants. Thus the four garden cities combined accommodate less than 100,000 persons. It is obvious that development on this scale cannot have any direct decentralising effect on the millions of inhabitants of the conurbations.

But, while the genuine garden cities have remained small, the "garden-city idea" has had a signal influence on town planning at large. As already mentioned, the first Town Planning Act, 1909, empowered local authorities to limit the number of buildings per acre. The practical effect of this stipulation was not great, owing to the small number of such town planning schemes. In 1914, however, a prominent leader of the garden-city movement, Mr. (later Sir) Raymond Unwin, was made Chief Town Planning Inspector on the Local Government Board (to become, after the last war, Director-General of Housing in the Ministry of Health). Mr. Unwin further served as a member of the Tudor Walters Committee, whose Report was largely based on his book *Town Planning in Practice*, first published in 1909. The Report endorses the policy laid down by the Local Government Board in its circular dated 18th March, 1918. "In connection with schemes intended to secure Government assistance, the aim should be to provide that in ordinary circumstances not more than twelve houses (or in agricultural areas, eight houses) should be placed on an acre of land." [1]

The opportunity of making this low density the general standard for suburban building arose after the last war, when practically no houses for the working classes could be erected without subsidies from public funds. The limitation of the number of houses to twelve per acre, which was introduced in 1918, has held the field ever since. "For many years," stated the Barlow Commission in 1939, "twelve to the acre has been the general upward limit applied to the density of subsidised cottages." [2] Unsubsidised building, it should be added, has equally complied with this standard.

[1] *Tudor Walters Report*, Para. 58. [2] *Report*, p. 64.

While the building of cottages, twelve to the acre, is the predominant rule, a number of block dwellings have been erected in the central districts of large towns. From 1933 onward, slum-clearance and abatement of overcrowding have been the main features of the Government's housing policy, entailing the re-development of many central sites. The Housing Act, 1935, in particular, provided for an Exchequer contribution towards the expenses of local authorities in erecting blocks of flats of not less than three storeys on expensive sites, i.e. in central districts. The subsidy is a graduated contribution per flat according to the cost of the site.[1] This policy of promoting compact building implies the recognition that at least for part of the working-class population central accommodation in blocks of flats is needed, namely, for the poorer sections of the community who cannot afford the high fares between outlying housing estates and central workplaces, and for some categories of workers such as railwaymen, dockers, market-helpers and night-workers who must live near their central workplaces. On the whole, however, the erection of blocks of flats is regarded as a bitter necessity, as an inferior solution compared with the ideal of urban development on garden-city lines. This is reflected in statements like that in the Barlow *Report* (p. 73) : " In existing circumstances a certain amount of flat construction for the working classes is unavoidable in large cities."

Yet there are also expressions of doubt whether the twelve-per-acre standard is the most suitable for urban growth. Sir Gwilym Gibbon states that

> " open development " is apt to be too mechanically applied, " twelve to the acre " (although the twelve may be ten or less) is almost an established article of faith. . . . In estate planning, for instance, it would probably have been better if there had been more close building of houses, with, on the whole, less open curtilage to individual houses and more land devoted to general open places and allotments.[2]

Other authors go further : they reject outright development on

[1] See Housing Bill (Bill 19, 1934), Financial and Explanatory Memorandum, p. i—The Housing Acts, 1930 and 1933, which provided for grants for slum-clearance, had made a start with paying subsidies towards the erection of high tenement buildings at higher rates than for other accommodation. See K. Liepmann, " English Housing Policy since the War " in *American Economic Review*, 1937, p. 503 ff.

[2] *Problems of Town and Country Planning*, 1937, pp. 110–11.

garden-city lines as the ideal model for town growth and advocate compactness as the appropriate manner of urban building. "Surely," says Thomas Sharp, "the ideal town is one which is as compact as the minimum requirements of public health will allow."[1] Elizabeth Denby holds that "the problem is, how best to tighten the spreading fabric of the towns, rehouse the people within the existing boundaries in planned humanised communities".[2] In recent years this doctrine has found more adherents.

ii THE RESULTS OF OPEN DEVELOPMENT EXAMINED

The policy of open development can be historically understood and to a certain degree justified as a reaction against the urban congestion prevailing in Victorian days. However, "the garden-city principle applied to suburbs" as a general device has not resulted in a satisfactory lay-out of large towns and has itself created new difficulties. The outcome of twenty years of twelve-to-the-acre policy has been the unwieldiness of London and other conurbations, of which the costs and hardships of the journey to work are a prominent symptom. The negative sides of this policy have come more and more to the fore; the reason is that the fundamental implications of low density of building have been ignored. These derive from the specific character of land as a thing which can be neither augmented nor removed. The failure to realise the intrinsic restrictions imposed by this monopolistic nature of land has led the promoters of open development to overestimate the applicability of their idea; they have administered a noxious overdose of a remedy which is no panacea but can cure only the complaints of a small portion of the town population. Relying indiscriminately on the technical facilities of transport, the low-density policy has overshot the mark: instead of a wholesome loosening of the urban fabric, there is now interminable diffusion[3] and much waste.

The Stock of Land Limited by Nature.—Concerned as it is with land as space, town planning faces a peculiar problem, not met with in efforts for the well-being of the people in other directions.

[1] *Town and Countryside*, 1932, p. 149.
[2] *Europe Re-housed*, 1938, p. 260.
[3] Prof. Abercrombie predicted such a result in 1924, in *Sheffield, a Civic Survey*, p. 72.

The progress of prosperity and the rise of the general standard of living usually takes the following course : goods and services which have so far been enjoyed only by the well-to-do become available to the masses of the people. This progress is, in principle, without difficulties, as far as it concerns goods and services which can be produced in plenty, such as agricultural produce, either home-grown or imported, and manufactured goods of all kinds. The same principle holds good with regard to sanitation, medical care and entertainment, which can all be expanded to cover the whole population ; and it is, subject to individual qualification, also true of education. Housing, however, because it involves the use of land, is a different case : "Land in an old country is a permanent and fixed stock." [1] Land cannot be amassed where it is most in demand. The space for urban building is restricted by distance from the centre ; there is a limit to the area that the heart of the town can make pulsate with vigorous life. "The quality of London life inevitably becomes attenuated as one recedes from the centre to the periphery." [2]

In former centuries, the fact that walking was the general form of locomotion restricted the urban area. Absence of public transport facilities kept the boundaries of towns narrow, even where this was not enforced by town walls. The majority of town-dwellers had, therefore, to be content with small plots. In the eighteenth century, a town house surrounded by a garden was the privilege of a few. Middle-class houses had not, as a rule, private gardens of any considerable size ; their occupiers, however, had access to parks and to semi-private "squares" which they shared with their neighbours, as is witnessed by estates in London, Bath and other towns, where the houses still remain as they were.[3] The masses of the people were crowded together in more or less congested streets.

The development of transport facilities in the nineteenth century made it possible for many middle-class families to move out from the centre to the fringes of the large towns and there to have private gardens at their doorsteps. By this adoption of a low building density for the residential suburbs, the built-up area

[1] A. Marshall, *Principles of Economics*, 8th Edition, p. 431.
[2] W. A. Robson, loc. cit., p. 440.
[3] At some time London could boast of some 460 domestic squares—see Thomas Sharp, *English Panorama*, p. 35.

was quite considerably enlarged; but the increase was as yet on such a scale as to cause a change in magnitude only, and not in kind: the area of the towns remained manageable.

This applies also to the garden cities proper, because they have remained small townships. The genuine garden-city principles couple low density of building with a limit as to the number of inhabitants: originally 30,000, now 50,000 is regarded as the upper limit. Actually Letchworth and Welwyn Garden City were, before the outbreak of the present war, far below those marks. Open development for a few thousand houses however (5,000 in Letchworth and 4,000 in Welwyn) has no serious consequences with regard to the scarcity of land in the neighbourhood of large towns. On the other hand, such a limitation of their population makes the garden cities entirely inadequate to serve as receptacles into which the inhabitants of the central districts might be decanted on any appreciable scale, and therefore inadequate as a means of relieving the congestion of the large cities.

Neglect of the Scarcity of Land.—The insufficiency, but also the harmlessness of the garden cities was abolished by the unorthodox measure of adopting the twelve-per-acre limit as the general standard of density, without also adopting the corrective stipulation of a limit in numbers. Town-planning policy has yielded to the temptation presented by modern transport, without giving heed to the penalties imposed by the dispersion of the urban population. Transport certainly widens the urban area, but it cannot do away with the basic quality of land as space and, therefore, as a restrictive element.

The brake of the limit in numbers of population being removed, the twelve-per-acre standard has, within twenty years, produced those vast built-up suburban areas which present so many difficulties to-day. A density of twelve as against forty to fifty houses per acre which had been allowed by the model by-laws of the later nineteenth century means that about four times the former acreage is required for accommodating the same number of families. Even now, when only part of the urban population has been housed and rehoused on the town fringes, "industrial and housing development . . . is stated to have covered with bricks and mortar an area equal in size to the counties of Buckingham and Bedford combined".[1] It is, how-

[1] Barlow *Report*, p. 15.

ever, an essential consideration that the built-up area should not cover more than necessary of the nation's land, in accordance with the aims to preserve the countryside and to minimise the alienation of good agricultural soil. This consideration is entirely lacking in suggestions like that of Sir Raymond Unwin that the land-owners may indemnify themselves against lower land-values by selling more acres for a given number of houses.

Consequences.—Even if sites for twelve houses to the acre should be shown to be cheaper than sites for compact building, that gain would be offset and heavily outweighed by the increase in the costs of all sorts of services in a loosely built area.[1]

As to the journey to work, low building density is certainly not the sole source of the travelling difficulties ; other factors are responsible for a large part of the costs in money, time and energy of travelling to work. But open development aggravates the difficulties. As suburbs or housing estates, developed at twelve houses per acre, occupy more land than if laid out compactly, the subsequent housing schemes are pushed farther away from the town centre. Those who live on the outskirts of a loosely built conurbation have, therefore, longer distances to travel to central workplaces than if they came from the fringes of a compact town with the same number of inhabitants.[2] Secondly, the average way from home to station is longer in a thinly spread than in a densely built district. It has been shown in the chapter on the Price of the Journey to Work, how much the walks to and from the station lengthen the total time spent on the journey. The use of a bus as a tributary of the railway, on the other hand, adds at least 2d. to the daily fares and adds, moreover, to the inconveniences of travelling through the changing of vehicles with all this implies. Finally, railway service for widely scattered suburbs is expensive : in spite of the long ways of access to the stations which the inhabitants of the suburbs accept, a station in such districts caters for fewer passengers than one in the densely occupied parts of the town ;

[1] Actually, " the municipal housing schemes in England at twelve houses to the acre have probably paid much the same price per acre as if they had put 18 or 20 houses on each, and have used 50% more land, to the great benefit of the land-owning class, as well as of the occupants ". Sir R. Unwin, Barlow Commission *Evidence*, p. 849.

[2] " The spreading of the population over a wide area with a density of twelve houses to the acre results in a lengthening of the route and requires an infinitely greater number of vehicles." S. Pilcher, *Road Passenger Transport*, 1937, p. 296.

this means increased costs of stations, which have to be maintained irrespective of the numbers who use them. To a certain extent it is possible to reduce the costs of a line by setting the stops wider apart ; but there is a limit imposed to this by the regard for efficient transport facilities. On the London Underground Railways, for example, about " mile spacing (as it exists at present) . . . would almost seem to be a limit for service ".[1]

Transport facilities are not the only service to be confronted with higher costs by low-density building, as compared with a more compact lay-out. Similar conditions prevail with respect to most other services. Open development means more street mileage for a given number of houses, entailing higher expenditure for maintenance, cleansing and lighting as well as for construction. The urgent stress with which the Tudor Walters Report insisted upon the uttermost reduction in the width of the streets and upon drastic simplification of their construction clearly indicates that the length of the streets implied in low density of building causes high expenditure. " The cost of the roads and sewers forms a heavy item in the cost of the cottage." [2] Longer streets involve, moreover, greater length, and thus higher cost, of drains, sewers, water, gas and electric-light mains, and of telephone wiring. It would be instructive to learn the additional costs caused to the public utility services by the twelve-per-acre standard, but these data have not so far been made known.[3] A greater mileage of streets entails further an extension of the

[1] Frank Pick, Barlow Commission *Evidence*, p. 358.

[2] Loc. cit., para. 74.

[3] Comparative studies have been made of the costs of street maintenance and cleansing, sewerage, refuse collection, public lighting and fire brigades in various towns and groups of towns according to size in numbers of population : C. Ashmore Baker " Population and Costs in Relation to City Management " in *Journal of the Royal Statistical Society*, 1910–11, p. 73 seq. ; Ministry of Health, Barlow Commission *Evidence*, p. 152 seq. ; H. S. Phillips, " Municipal Efficiency and Town Size " in *Journal of the Town Planning Institute*, March–June 1942, p. 139 seq. All three publications mention the towns spatial size as influencing the costs of these services. C. A. Baker showed even a remarkable general correspondence between the density (area in square miles per 1,000 persons) on the one hand, and the costs of gas and electricity on the other hand. The Ministry of Health, in its more recent documentation, mentions mileage and width of the streets among the factors on which the costs of cleansing and so forth depend. But all tables give simply ratios of costs per head of the population. The results of the comparisons are inconclusive ; even H. S. Phillips, whose investigation seems to point to an optimum technical position of towns somewhere between 100,000 and 250,000 inhabitants, states that " the correlations between . . . services and size . . . are mostly insignificant ". It may well be that by taking into account the length and width of the streets served or used by the services under review, definite correlations between costs and " size " of town would be made visible.

policeman's beat and longer routes for the postman and the dustman, thus raising the costs of these services. Private business is not exempt from this burden : the newspaper boy, the milk roundsman, the coal and laundry vans, all shops which deliver their goods to the customers, the doctor visiting his patients, the housewife on her way to the shops—in fact, every citizen on every private and business errand, pays for low density of building with loss of time, fatigue or expenses for transport in making his way through the lengthy streets. It is not possible to assess the waste of time and energy imposed on the inhabitants, and little is known of the monetary costs—these are borne by the tax- or rate-payers, charged to the consumers or are reducing the incomes of the tradesmen. The results of the one pertinent inquiry, into the distribution of milk, are revealing.

> As building density rose from 250 to 350 houses per mile of street, there was a fall in weekly delivery costs per household of 9·3% in congested areas and of 7·7% on housing estates.[1]

Also

> delivery costs are higher on housing estates than in congested areas . . . partly owing to the lower building density and . . . to the greater distance of the houses from the street and the practice of backdoor delivery on housing estates . . . The size of a town, measured not only by its population, but also by the area within its boundaries, clearly has a direct bearing upon delivery costs.[2]

To these disadvantages must be added the difficulties, met in districts where the urban fabric is thin, of providing statutory services and of developing a community sense and public spirit in general. It is obvious that the total burden of these drawbacks must weigh heavily against any possible advantages of low building density.

iii THE CASE FOR MODERATE COMPACTNESS

The outcome of the foregoing discussion is a seeming incompatibility of the hygienic and the economic requirements with regard to density of building : a reduction of urban congestion is highly desirable in the interests of public health and, it is claimed, also for reasons of beauty and convenience. From

[1] John Cripps, *The Distribution of Milk.* A Study of Town Delivery Costs. Oxford, 1938, p. 75.

[2] Ibid., p. 89.

the economic and certain social points of view, on the other hand, low density of building has very considerable disadvantages. The task in hand is to find a mid-course, that is to say, the highest degree of building density which is compatible with modern standards of public health.

Those inclined to favour open development have not been unaware of the importance of this question. Professor Abercrombie said in 1933 : " The precise relation of density to health has always been difficult to ascertain." [1] Even an author who in later days stigmatised a density of twenty houses per acre as " overcrowding " was originally less dogmatic : " It is not possible to fix any absolute limit for the number of houses to the acre which can be regarded as a maximum compatible with health and comfort . . . It is not easy yet to weigh the disadvantages that might arise from enlarging our towns to such an extent as would give a much lower number of houses to the acre all through." [2]

The question of the adequate density of building can be tackled from various angles.

The Relation of Health to Density.—As the Barlow Report (p. 156) points out, it is essential to distinguish between mere concentration of population and congestion : it is the latter which is harmful. High density of building, however, is only one of several constituents of congestion, the others being overcrowding of persons per room and doubling-up (or even trebling or more) of families in private dwelling houses. By abatement of overcrowding and provision of self-contained dwellings for individual families, together with sanitary improvements, " remarkable advance . . . has been made in overcoming and eliminating the disadvantageous elements which formerly influenced the public health in large towns ".[3] This statement acknowledges the obvious fact that other factors besides building density determine public health. Logical reasoning on the relation between health and density would, therefore, proceed to ask : after such improve-

[1] *Town and Country Planning*, p. 107.

[2] R. Unwin, *Town Planning in Practice*, 6th impression, 1919, p. 319. On p. 320 Sir Raymond went on to say : " That a greater number of houses to the acre than 12 may be planned and yet produce a healthy suburb is proved on the estate of the Ealing Tenants and many others. On some parts of the estate at Ealing the houses approach 20 to the acre, excluding roads and larger open spaces reserved for recreation grounds."

[3] Barlow *Report*, p. 58.

ments have been achieved without greatly reducing the number of dwellings per acre in the inner town, how much has this number to be lowered in order to arrive at a satisfactory urban density? Is the limitation to twelve houses per acre actually necessary for safeguarding health? Unfortunately, this crucial question has not been raised by the Barlow Commission; the twelve-per-acre standard is taken as an axiom.

An attempt to prove that compact building adversely influences health is made by comparing mortality in urban and rural districts.[1] But firstly, it is not justifiable to attribute higher death rates in large towns to pure density considerations; secondly, the medical evidence brought before the Commission is inconclusive. It does not invalidate the view that, if houses are decent and overcrowding is prevented, there is no reason to think that the inhabitants of a compact town would be less healthy than those of rural districts.[2] What is required is not a transfer of rural densities to the town, but the adaptation of the benefits of country life to urban environment. "The great need of a town population," says the Ministry of Health, "is access to the country or the equivalent of what the country means—fresh air, sunlight, facilities for exercise and recreation and means for the harmless and profitable employment of leisure."[3] By way of proper planning, however, none of these requirements is incompatible with moderately high density of building.

Utilisation of the Individual Plot.—Another approach to the problem of the optimum density is to examine the size of the individual plot and the different parts into which it is divided. Do the functions ascribed to these parts justify a plot of one-twelfth of an acre? Apart from the space taken up by the house, there are, under open-development regulations, the back garden, the front garden (or forecourt) and the interspace caused by the detachment of the houses.

The back garden is the largest item. Town planning literature provides one attempt to justify the size given to this main constituent of the plot: "Twelve houses to the net acre of building land, excluding all roads, has been proved to be about the right number to give gardens of sufficient size to be of com-

[1] Ibid., pp. 56–63.

[2] H. S. Phillips, loc. cit., p. 143, comes to the conclusion that "size is a neutral factor in towns as far as health is concerned."

[3] Ibid., p. 65.

mercial value to the tenants—large enough, that is, to be worth cultivating seriously for the sake of profits, and not too large to be worked by an ordinary labourer and his family." [1] This is obviously no vindication of the twelve-per-acre standard. The commercial gain by the individual gardener does surely not set off the disadvantages to the community of suburban spread, supposing even that this gain is not outweighed by his and his family's own sacrifices for long daily journeys. It must further be doubted whether the underlying assumption is right, that each family—not to speak of the single members who are expected to lend a hand in gardening—is inclined and able to cultivate a garden. This doubt is reflected in the following discussion before the Barlow Commission :

> " In any population the tastes of the people with regard to gardening are by no means uniform, and with a regular density of twelve houses to the acre some garden-plots might not be fully utilised by the inhabitants ? " " Yes . . . We suggest it is better for them to live near their work rather than that everyone be compelled to have a garden whether they want one or not." [2]

The problem of providing private gardens to urban houses becomes easier of solution if a distinction is made between house gardens for pleasure and allotments for food production. Most people like to have a garden with flowers, but not everybody wants to grow vegetables.[3] The raising of vegetable crops, whether for sale or for domestic consumption, is to be regarded as part of agriculture ; the place of the allotment is not within the built-up urban area, but in the green belt surrounding it. The allotment is that form of garden to which Wren's maxim should be applied : "All churchyards, gardens and unnecessary vacuities . . . to be placed out of the Town." [4] Gardens proper, on the other hand, should be, and before the war largely were, only for pleasure, and can be small.[5] The greater compactness of building which thereby becomes possible facilitates the access to the allotments beyond the built-up area, for those who care for their cultivation.

[1] R. Unwin, loc. cit., p. 320.
[2] *Evidence*, p. 616. Hundred New Towns Association.
[3] This discussion is concerned with peace-time conditions, and not with the present emergency.
[4] *Parentalia*, edited by Stephen Wren, 1750. Quoted from *Christopher Wren.* A bicentenary memorial Volume, 1923, p. 169.
[5] This was realised by H. R. Aldridge : " The home garden . . . may not be more than a few square yards ". Loc. cit., p. 140.

The small front garden (and also the side space resulting from the detachment or semi-detachment of the houses) is meant to give each cottage privacy and social and architectural distinction. While a proper device in busy traffic roads, the forecourts would seem to be out of place in quiet residential streets with small two-storey houses.

Of the arguments for the space taken up by detachment, the one already mentioned, for the very sake of detachment, carries little weight in face of the importance of economising frontage. Nor is detachment recommended by " the violent draught which often blows through the narrow spaces between pairs or groups of houses ".[1]

From the economic standpoint, both front garden and detachment are objectionable, as they cause, without proper justification, increased consumption of building land.

Thomas Sharp strongly attacks both devices of open development from the point of view of beauty ; through the deep building line the " towns have lost all scale and dignity ", and detachment has a pernicious effect on harmony and architectural composition of the street.[2]

The hopes of the open-development movement to create beauty have been frustrated. " We are beginning to see that the new garden suburb, even when well designed, if extended illimitably can become monotonous . . . and the speculative builder's version of low density, interminable semi-detached villas, has already achieved weariness in no uncertain sense." [3]

3 *CONCLUSION*

The policy, pursued in the inter-war period, of open development of the suburbs has been an endeavour to house the urban population in rural surroundings. But the experiment teaches the lesson that a very high price has to be paid for the attempt to have the best of both worlds. It is generally realised that the scene of the town-dweller's daily life cannot have the natural beauty of the countryside. It must equally be accepted as in the

[1] *Tudor Walters Report*, para. 67.
[2] *Town and Countryside*, p. 157 and p. 154.
[3] P. Abercrombie, *Town and Country Planning*, p. 108.

nature of the circumstances that the masses who work in the centres of large towns cannot have both, nearness between home and work, and private gardens at their doorsteps. So far, there have been two different ways of living for the urban population : " living in " which means to forego private gardens, and " living out " which means as a rule to put up with journeys to work. Each way suits different people, according to type of family, kind of work and individual inclination.

Resignation to the incompatibility of these two advantages is made easier by the realisation that the chief points raised against compactness of urban building are unsubstantial under modern conditions. The development of sanitation in particular makes the arguments against compact building much less valid than formerly. Nor need higher building density mean absence of light and air and open space. Terrace houses with small private gardens are one solution. Blocks of flats can be surrounded by communal gardens and playgrounds ; the privacy and the community life afforded in existing blocks of flats can both be improved.[1] Provided the built-up area does not stretch too far out, access to the open country can be made easy and pleasant by green wedges penetrating into the heart of the town, as devised in Mr. Pepler's Park Diagram.[2]

The need for easy access to the countryside from the central districts is one of the main arguments against low density of building even on the fringe of large towns. Nucleated development on a moderately compact plan is required also in the interest of the suburban population ; it contributes to social cohesion, which does not thrive where the urban fabric is loose ; it makes for economy in most services, and it shortens the journey to work. Rejection of the twelve-per-acre standard for the suburbs does not mean abandonment of private gardens. While front-gardens and detachment should be abolished and back-gardens reduced in size, small flower-gardens adjoining the houses are to be provided and allotments to be made available in the surrounding green-belt for those who care to raise their own vegetables.

There is, however, one feature of recent development which

[1] See Elizabeth Denby, *Europe Rehoused*, p. 263 seq. ; also *When We Build Again*, pp. 116–17.

[2] " Open Spaces " in *Town Planning Review*, 1923, p. 16.

allows part of the industrial population to have private gardens and yet to live near their work, although not in the inner town. The movement of factories outward from the town centre was a marked trend between the two wars. It should therefore become possible to locate light industries on planned sites among the residential areas on the outskirts of towns, thus keeping the decentralisation of industry "commensurate with the type and size of housing estates in the neighbourhood".[1]

Nucleated development of the outskirts of conurbations includes satellite towns of mixed residential and industrial character. It must be kept in mind, however, that this form of urban settlement can benefit but a restricted part of the industrial population: a satellite town is the proper place only for plants of moderate size. Large-scale undertakings, frequently demanded by technical and economic considerations, need the labour supply of a large town or of many smaller places to draw upon. Location of a big works in a satellite town would involve it in the dangers of a one-industry town.

If attention is focused on the correlation between home and workplace, there appear four combinations of urban lay-out; with moderate compactness of building all of them can be justified by social, economic and technical considerations.

(*a*) "Live in and work in", i.e. homes situated near workplaces in the centre of the town (some secondary earners may have to travel to work in other districts). The advantages are obvious, provided housing conditions are decent and access to the open country is easy.

(*b*) "Live out and work in", i.e. severance of dormitories and exclusive workplaces. This is in line with the general tendency of a relaxation of bonds and of specialisation in the ways of living, on the condition that the social machinery is adjusted to the dualism of places and that the building density is not too low.

(*c*) "Live out and work out", i.e. satellite towns which provide employment for the bulk of their earning inhabitants. Such circumscribed urban units, developed on a plan of moderate compactness, would relieve the pressure and unwieldiness of the

[1] See W. G. Holford, "Notes on Segregation in Industry" and "The Location and Design of Trading Estates" in *Journal of the Town Planning Institute*, 1938, p. 234 ff., and 1939, p. 158 ff. respectively.

central town and yet enable the residents of the satellites to share the opportunities of the metropolis.

A second pattern of living out and working out is the grouping of several dormitories within a convenient distance of a trading estate. Further experience and research will have to show what proportion of urban workers can find employment in such moderately sized industrial zones.

(*d*) " Live in and work out ", i.e. a new form of urban development, originating from the location of factories on the fringes of the town, while employees continue to live in more or less central quarters. This tendency is in the initial stages, and it cannot yet be judged how far it will go. The people are kept in propinquity to each other and near the foci of social and cultural life ; daily travelling gives them access to various outlying industrial zones.[1]

Each of the four solutions meets different requirements of modern society. The object of town planning must be to blend these various types of lay-out in such a manner as will best serve the multifarious purposes of the community and of the individuals and families of which the community is composed.

[1] This arrangement resembles the situation of the *Ackerbürger* of a former age, that is of the farming townsman, who had to leave the precincts of the town on his daily way to work on the farm.

PART B: Statistical Inquiries

Chapter I

PREVIOUS INVESTIGATIONS

THE many aspects of the journey to work hardly lend themselves to examination by a single statistical inquiry. Each source of information allows only certain features to be studied. The data provided by a Census reveal the demographic conditions relevant to the daily movements, but fail to throw light on the journey. Transport statistics, on the contrary, usually yield no other information than a few data on the journey; in one exceptional case only, railway records furnished the material for a comprehensive investigation. Monographs which are based on the personnel records of individual firms provide evidence of the effects of daily travelling on the workers' health and efficiency but neglect the demographic aspect.

For the following review of previous investigations, the more important publications have been selected from a number of others which give less comprehensive information or proceed on the same lines as those reviewed.[1]

I *INVESTIGATIONS BASED ON A CENSUS OR SIMILAR MATERIAL*

a GERMAN CENSUS, 1900

The German Census in 1900 was the first statistical attempt to find out the earners who work outside their places of residence: the Census schedule included a question as to the workplace of all earners in the family. Several of the Federal States took the matter up and tried to systematise the results, first and foremost Wurttemberg;[2] the whole State was surveyed as to the inter-

[1] For a bibliography see the report by the International Statistical Institute reviewed on p. 129 fol.

[2] *Württembergische Jahrbücher für Statistik und Landeskunde*, 1902, p. 237 seq.

connection of places by daily journeys to work. Dr. H. J. Losch, who coined the term "Pendelwanderung", stressed the importance of this new branch of statistics. The main results were the following: an unexpectedly high number of daily journeys was revealed; about one-tenth of all gainfully employed persons (excluding agriculture and domestic service) worked in places other than their areas of residence. Nearly all of them were employed in large undertakings. One-quarter of the daily travellers were females. The capital, Stuttgart, with some 177,000 inhabitants, attracted 8,500 earners from outside. Many cross-currents and counter-currents were disclosed, in addition to the main streams towards the important labour markets. The inquiry was repeated at the following censuses; in 1925, one fourth of Wurttemberg's industrial workers were employed in localities other than their places of domicile.

The Prussian examination of the data of the 1900 Census [1] was less comprehensive but illuminating in some respects. The investigation was restricted to towns of 100,000 or more inhabitants and to certain industrial districts. It was not possible, therefore, to show the manifold interconnection by daily journeys of a great number of smaller places; even where included, the smaller places were treated only as feeding the nearest big labour market. On the other hand, the attempt was made to classify the residential places according to their location relative to the one large town: (*a*) suburbs more or less ripe for incorporation; (*b*) places farther off but falling within the pull of the large labour market; (*c*) places at a greater distance, connected with the town by railway. This classification did not, however, prove satisfactory, and adoption of a zoning by kilometres in further investigations was suggested.

In Hamburg, the Censuses 1900, 1910 and 1925 were used to find out the daily movements between various districts of the town; it could thus be shown how far the transformation of the town centre into a mere business district had gone.[2]

b Swiss Census, 1910

In Switzerland, a relevant investigation was made at the 1910 Census.[3] Emphasis was laid on the marital conditions of

[1] *Zeitschrift des Preussischen Statistischen Landesamts*, 1904, p. 1 seq.
[2] *Statistik des Hamburgischen Staates*, 1902, p. 87 seq.
[3] *Schweizerische Statistische Mitteilungen*, 1919, p. 1 seq.

the daily travellers, in view of the fact that social services for these persons and their families have to be provided at their domiciles, while the workplaces have no such financial burdens on their behalf.

c CENSUS OF ENGLAND AND WALES, 1921

i REGISTRAR-GENERAL'S REPORT

In England and Wales the 1921 Census was used for a country-wide survey of *Workplaces.* The results were presented in the following way.

Table 1 classifies the occupied population of each sex enumerated in each local government area according to the description of workplace : total population ; total occupied residents with workplace (*a*) in the area, (*b*) outside the area ; these figures are supplemented by persons working in the area but enumerated elsewhere.

Table 2 shows the places between which the daily journeys are actually made : (*a*) number of persons enumerated in a given place A but working elsewhere, viz. in B, C, and so forth ; (*b*) working in A but enumerated elsewhere, viz. in Q, R, and so on. Movement from or to individual areas is only shown when 25 or more persons are involved (for London and five Home Counties, 20 or more).

Table 3 repeats the enumerated, that is the night population, and then shows the day population, which is found by deducting the outflow from, and adding the influx to, the enumerated population. Next come the aggregate daily in-and-outward movements of each area, numbers as well as ratios of enumerated population ; the final column shows the net outward or inward daily movements, again by number and by ratio. The arrangement of the tables in *Workplaces in London and Five Home Counties* is slightly different, and two additional tables classify the population by zone of enumeration and by zone of workplace ; the places are arranged in four zones, according to their distance from Charing Cross (p. vii). The result is summarised in the table on page 114.

In most areas in the Home Counties and in 16 Metropolitan Boroughs the net movement is outward ; the day population is concentrated towards the centre of the region. Apart from the

City with a day population of 436,000 or over thirty times its night population, Westminster, Holborn, Finsbury and St. Marylebone also show large increases during the day.

DISTRIBUTION OF OCCUPIED POPULATION ENUMERATED IN LONDON AND FIVE HOME COUNTIES

Zone (distances from Charing Cross)	Distribution by Zone of Enumeration		Distribution by Zone of workplace		Excess (or deficiency) of enumerated over working population
	Number	%	Number	%	
London (Admin. County)	2,173,691	49·1	2,618,392	59·1	−444,701
Outside London Admin. County but within 10 miles circle	1,190,097	26·9	786,279	17·8	403,818
Outside 10 but within 15 miles circle . . .	144,396	3·3	135,271	3·1	9,125
Outside 15 miles circle but within Home Counties	919,337	20·7	820,976	18·5	98,361
Outside Home Counties [a]	[a]	[a]	66,603	1·5	[a]
Total (enumerated in London and Home Counties)	4,427,521	100·0	4,427,521	100·0	[a]

[a] The London volume was published before the enumeration outside the Home Counties was finished. Some 65,000 persons, mostly holiday-makers, who were enumerated outside the Home Counties gave London as their workplace.

Reference is made to the multifarious aspects of "this new factor in demography"; further to the many difficult problems of traffic, transport and housing on which the daily movements have a direct bearing; the environmental influence of the workplace, often entirely different from the worker's residence, is mentioned, and so are the administrative problems arising out of the increasing "divorce" between residence and workplace.

The call for research implied in these remarks has not met with much response. And in the following Census, in 1931, the question asking for the workplace was discarded from the schedule.

ii MONOGRAPH ON THE MERSEYSIDE BY W. HEWITT

What good use can be made of Census returns for throwing light on local conditions concerning the journey to work may be seen from a study of *Workplace and Movement of Workers in the Merseyside Area*, by W. Hewitt, 1928.

Mr. Hewitt uses the workplace data of the 1921 Census to examine the areas from which each of the four Merseyside boroughs, Liverpool, Bootle, Birkenhead and Wallasey, draws its labour. He groups the places of residence geographically and by main means of transport (railway lines—traffic across the estuary of the Mersey). The interchange of workers between the four adjacent boroughs surpasses the influx from elsewhere, and the study is mainly concerned with their interconnection by the journey to work.

The author combines the workplace results with such statistical information as is available on the occupational and industrial character of the various places.

The effect of the general industrial character of the workplace on the composition of the daily movements is clearly discernible in Bootle, a town offering occupation mainly to men : a net daily influx of males goes with a considerable exodus of women and girls. The same tendency, though less marked, is apparent in Birkenhead. Wallasey, a predominantly residential place, has half its male earners working outside.

There follows a brief survey of other towns in S.W. Lancashire and in West Cheshire, mentioning the strong counter-currents of daily workers to and from a given place. The fact that workers had not moved nearer their work is partly attributed to the housing shortage prevailing after the last war. The analysis might have been carried a little further by consulting Table 1 of the Census Report as to the ratio of females among the daily outflow from certain places : girls often prefer longish journeys to giving up their parents' home for lodgings near their work. Another aspect which might have been enlarged upon is the position of small places which serve as dormitories for the main labour markets. What proportion of their outgoing earners are attracted by the various workplaces within daily reach, in other words, where and to what extent does the pull of various industrial centres overlap ? [1]

Among the various problems involved in the daily movements,

[1] To give an example : 113 persons coming to work in Liverpool from Little Crosby represent 64% of the outflow of this place ; on the other hand, the 122 residents of Wigan C.B. working in Liverpool account for less than 1% of Wigan's total overflow.

this study recognises in particular "the resulting division of personal interest, on the part of this tidal population, between the working place on the one hand, and the place of residence on the other—neither locality evoking the whole-hearted interest of the man or woman in its good government, its amenities and its general progress ".[1]

d Special Investigation in the Industrial Region of Central Germany, 1929

An attempt to combine demographic comprehensiveness with a measurement of the journeys was made in an investigation of the daily movements in the Industrial Region of Central Germany,[2] an area of rapid industrial development. The enumeration was based on local government areas which involved disregarding the distance between the places of residence and of work. Some measurement of the journeys was, however, provided from other sources, and by a schematic method it was possible roughly to classify the journeys, particularly to distinguish short walks, maybe just crossing a street into another local government area, from genuine journeys involving considerable time, strain and costs.

The results of general interest were the following. More than every tenth of the gainfully employed persons in the survey region were found to work outside their places of residence. Centres of attraction were: the large towns (Leipzig, Magdeburg, Halle); the big chemical plants, erected only after the last war in the open country, foremost the Leuna Works;[3] the bituminous coal-pits. The daily movements were fed by the rural population and also by mobile inhabitants of all areas, including those which were themselves important workplaces.

The majority of places sent out less than 50 persons each, every morning; they accounted for only one-eighth of the total daily movements. On the other hand, nearly one-third of

[1] Loc. cit., p. 4.

[2] " Die Pendelwanderung im Mitteldeutschen Industriegebiet " in *Vierteljahrshefte zur Statistik des Deutschen Reichs*, 1931, I. The inquiry was carried out by the Statistische Reichsamt; the present writer was closely associated with it.—The basic data were obtained from the income-tax authorities which used to take annual censuses for official use only. Their schedule was similar to that of a public Census, but included a question as to the earners' workplaces, for the purpose of allocating certain grants-in-aid: the yield of a tax on total wages paid by a firm (" Lohnsummensteuer ") was proportionally redistributed to the workers' places of domicile.

[3] The correct name is " Ammoniakwerk Merseburg ", but it is rarely used.

all travellers came from places sending out 1,000 or more. The conflux at workplaces was remarkable; six of them had a daily intake of at least 5,000. Outstanding among these places were Leipzig and Leuna. While, however, the influx into Leipzig amounted only to about 4% of its 700,000 inhabitants, Leuna's day population was many times as high as the night population; the reason is that Leuna, with about 22,000 employees, was a new plant, located among three or four isolated villages. Half of those who peopled Leuna during the working hours lived in the towns Halle, Merseburg or Weissenfels; the other half in 536 different areas.

The places under 5,000 inhabitants had an aggregate daily outflow amounting to nearly double their influx; the large towns, on the other hand, had an influx three to four times as high as the outflow. If Leuna is excluded, the localities up to 20,000 inhabitants had an aggregate excess of outward journeys, and only the towns over 20,000 an inward surplus.

Besides the main streams, there were cross-currents and counter-currents of considerable magnitude; one-half of all places under review had both in- and out-flow. Another remarkable result was the dispersal of the residents of single localities to various workplaces. An outflow of 30, 20, or even 10 earners sometimes dispersed to half a dozen workplaces; especially in the area between Leipzig, Halle and Leuna the pull of all three big labour markets was felt; even of the dormitory suburbs, only a few sent their outflow of earners in solid batches to the one nearest town.

As to the duration of the daily journeys, the main difference was between old-established and new works. While the former usually had the bulk of their employees, at least the stable elements, living within easy reach of the works, new factories, and especially those erected in the open country, were dependent on recruiting far afield. A striking contrast was presented by the Leuna Works, with 20% of its huge invading labour force travelling one hour or longer each way, and the mining districts where journeys of such duration were quite exceptional. A significant result of the inquiry was the highly complex and varied nature of the daily movements.

HOUSING CENSUS IN AMERICAN TOWNS, 1934

In the U.S.A., a *Real Property Inventory* [1] was used to obtain information on the duration of the daily journey and on the method of transportation. The survey covered 64 towns selected from all States so as to give a representative picture of urban housing in America. The towns had an aggregate population of over nine million and a total of some 2,600,000 families. In size, they ranged from under 12,000 to over a million inhabitants. Furthermore, New York, Pittsburgh and several other large towns took censuses on the same lines on their own.

The two relevant questions were : "Time required to get to Work" and "Mode of Transportation". Data were obtained only for the principal income-earner of each family, and only the means of transport chiefly used was listed. Secondary earners and subsidiary means of transport were thus left out of consideration.

The most striking result from the point of view of the present study is the outstanding importance of the private car for travelling to work in America. In 14 of the 64 cities under review, over one half up to nearly two-thirds of all principal earners used private cars to get to work ; in 38 towns, 30 to 50% of these men travelled by car, in 11 towns 20 to 30%, and only in one town (where 70% lived within walking distance) there were less than 20% car users.[2]

Walking was the other main method of transport : in 21 of the towns under review, one half or more of the principal earners walked to their workplace. As would be expected, high proportions of those who walked were found more often in small towns than in large.

However, the more or less frequent use of private cars seems to be related to the geographical region rather than to the size of the towns. In all six New England towns covered by the Real Property Inventory the private car was the method of transport for only 20 to 30% of the principal earners ; [2] in the Middle Atlantic towns for 22 to 40% ; at the other end of the scale are the Pacific towns where 49 to 63% of the principal earners used motor-cars for the journey to work, and the Mountain towns (with one exception) with 37 to 61% car-users.

[1] Department of Commerce. Bureau of Foreign and Domestic Commerce, Washington, D.C., 1934.

[2] Excluding those working at home.

In view of the modest rôle played in most of these towns by public means of transport, it is no serious shortcoming that the changing of vehicles was not recorded. The neglect of the daily journeys of secondary earners is a greater defect. To what extent do earning wives and children share a car either with the householder or with another person, on their journey to work; or work within walking distance from home; or use public means of transport?

For the larger towns, not only the "City Proper" was examined but also the "Environs", that is the adjoining areas sharing in the social and economic activities of the city. In most of these "Metropolitan Districts",[1] the Environs show a much higher proportion of car-users than the City Proper. In some cases, however, factories were obviously located on the outskirts, thus giving the suburbians the benefit of short daily journeys, often within walking distance.

It is significant that cycling is not listed among the specified modes of transport, but included with "other modes" which account for negligible percentages only. This is in striking contrast with the wide use of the bicycle in this country, as will be seen from a comparison with the statistical information collected for the present investigation.[2]

Information which such a Census can provide on the time required to get to work is likely to be approximate rather than accurate. Even so the results are interesting. Journeys of half an hour or more were made—

In 6 towns by under 10% of the principal earners
„ 25 „ „ 10—under 20% of the principal earners
„ 26 „ „ 20— „ 30% „ „ „ „
„ 6 „ „ 30 or more % „ „ „ „ (1 blank)

As one would expect, the frequency of long journeys increases with increasing size of the town; five of the six towns where journeys of half an hour or over were most frequent had over 300,000 inhabitants.

It is instructive to supplement these findings with those of the Real Property Inventory of New York, equally made in 1934.

[1] "Metropolitan Districts" have an aggregate population of 100,000 or more, and contain one or more central cities of 50,000 or more inhabitants.

[2] See p. 131.

As the following table shows, in three of the five boroughs 30 to 34% of the principal earners had journeys of a whole hour or longer. In some parishes in Richmond as many as 80% of the principal earners had to travel at least one hour each way.

DURATION OF THE JOURNEY TO WORK IN NEW YORK

Principal Income Earners only. 1934

Time required to get to work in minutes	Percentages of Total Reporting Families					Total New York City
	Manhattan	Bronx	Brooklyn	Queens	Richmond	
0–19 . . .	31	15	20	19	28	22
20–39 . . .	49	28	38	27	32	35
40–59 . . .	15	27	24	20	7	19
60 and more . .	5	30	18	34	33	24
Total population (000's) . . .	1,724	1,295	2,550	1,144	160	6,874

* Real Property Inventory of New York City, 1934, p. B.

f SAMPLE INQUIRY IN BIRMINGHAM, 1937–8

The study of the journey to work was one of the principal objects of the Bournville Village Trust's investigation of conditions of living and working in Birmingham before the war.[1] Their survey covers over 7,000 working-class families, being a sample of one in 35. Information is provided both on demographic factors and on features of the daily journey.

The City of Birmingham was divided into seven districts, one comprising the Central Wards, the others radiating outwards. The different situation with regard to the journey to work is reflected in the proportion of residents in each district who also work in that district. Apart from the Central Wards, the ratio ranges from 53·6% in the South-Western to 22·8% in the North-Western district. Cross- and counter-currents of the inhabitants of the various districts on their daily journeys are manifest. The Central Wards, for instance, comprise only 18·1% of the total inhabitants of Birmingham but provide work for 35·8%, to which must be added those who live outside the city boundaries and

[1] *When We Build Again*, 1941, pp. 62–77. The book was published after the present study had been completed.

travel in daily to work; at the same time, more than one-third of the inhabitants of the Central Wards travel out each day to work in other districts.

If, instead of the seven districts, three concentric zones—Central Wards, Middle Ring and Outer Ring—are compared, the different incidence of the journey to work stands even more in relief. According to expectation, the trend is towards longer journeys for those living in the Outer Ring (see the upper part of the following table). This is of great consequence, because more than half of the total population of Birmingham had come to live in the Outer Ring before the outbreak of this war. The municipal housing estates, in particular, are for the most part on the outskirts. Those living on new estates therefore travel farther to work. Of the municipal householders, 23·9% have journeys of four miles or over each way, of all householders, only 12·2%. Correspondingly, the municipal tenants are burdened with higher fares.

PERCENTAGES OF WAGE-EARNERS TRAVELLING . . . MILES TO WORK

Residential Zones Type of earners	Nil or under 1	1–2	2–3	3–4	4–5	5–6	6 and over	Vari-able
All Earners in								
Central Wards	58·0	21·9	6·8	3·4	1·3	·2	1·0	7·3
Middle Ring .	38·0	30·3	15·2	5·2	1·9	·5	2·0	6·8
Outer Ring .	21·0	20·0	17·5	16·6	10·8	3·7	2·9	7·6
Entire City . .	33·2	23·5	14·7	10·7	6·3	2·1	2·2	7·3
Principal Earners only . . .	28·9	22·4	14·2	11·5	6·8	2·4	3·0	10·9

In the lower part of the table the distances between homes and workplaces are shown separately for principal wage-earners and for all wage-earners, i.e. including secondary ones. It appears that among the sample families subsidiary wage-earners travel rather shorter distances than the householders.[1] The explanation is that a large proportion of the subsidiary earners consists of young people, including many girls, who find jobs, mostly dead-end ones, near home.

[1] The actual differences are greater than the figures indicate because principal earners are included in the totals.

The proportion of principal earners who go home for lunch is as high as 26·9% of the samples in the entire city; the position in the three zones is 34·9% in the Central Wards, 30·2% in the Middle Ring, and 22.5 % in the Outer Ring.

The presentation of the results of the survey has been restricted through the exigencies of war. It is to be hoped that the valuable material collected by this inquiry will be fully published when conditions permit. It would be particularly interesting to get information on the incidence of the daily journey on various families and on various members of the family.

2 *INVESTIGATION BASED ON TRANSPORT STATISTICS:* STUDY ON BELGIUM BY E. MAHAIM, 1910

The records of railway undertakings yield the numbers boarding and leaving the trains at the various stations, from which distances travelled and amounts of fares paid can be compared; but usually they neither furnish information on demographic conditions, nor show particulars of the passengers. A unique combination of circumstances prevailing in Belgium at the beginning of this century enabled Ernest Mahaim to produce a study both comprehensive and detailed.[1] At that time the train still accounted for the bulk of the journeys; the motor bus was almost non-existent and the use of the bicycle had not nearly reached its present volume. Trams, it is true, were working, but rarely over long distances. For manual labour, therefore, the workmen's train was practically the only means of transport for travelling to work, and an enumeration of workmen's tickets gave a fairly complete picture of the daily movements. Craftsmen and salaried employees in Belgium were (and are) excluded from using the specially reduced workmen's tickets. This restriction was the reason for particulars of the travelling workers being available; in order to ensure that only bona-fide manual workers get the benefit of reduced fares, the stationmasters ask for detailed data from each applicant for the weekly workmen's tickets; sex, age, occupation, home address and address of employer—just the items which, together with the data of the

[1] *Les Abonnements d'Ouvriers sur les Lignes de Chemins de Fer Belges et leurs Effets Sociaux,* 1910. The data refer to 1907 and 1908.

journey, are needed for a thorough inquiry into the daily movements.

Mahaim's results were the following: the use of workmen's tickets had almost turned the whole of Belgium into one labour market. Workmen's tickets accounted for 42% of all train journeys in Belgium; approximately one-fourth of the total industrial workers participated in them. The towns had an out-flow of skilled, the countryside of unskilled workers. The average length of the return journey was 19 km. (11·8 miles), excluding the approach from home to station and from station to actual workplace and back. The bulk of the travellers spent less than three-quarters of an hour in the train, morning and evening rides taken together. Some miners, however, had to travel up to 2¼ hours from door to door, twice daily. Mahaim formed six groups of industries and found that they accounted for the following percentages of the journeys:

Mining	Factories	Building	Navvies	Others	Railway administr.
10·0	30·1	14·9	15·6	19·0	10·4

The percentage of the miners is remarkable; they participated in daily travel in the same proportion as industrial workers.[1]

Mahaim was the first to recognise the social importance of the daily movements. He discussed their bearing on the rural exodus, on the labour market, on family life, on the moral and intellectual development of the people. While admitting certain disadvantages, Mahaim took, on the whole, a strongly favourable view of the journey to work as furthering social and economic progress. Although some of his opinions are no longer apposite, later writers are indebted to Mahaim for the broad view which he took in assessing the social significance of the journey to work.

3 *INVESTIGATIONS BASED ON FIRMS' PERSONNEL RECORDS*

Inquiries based on the material of a single factory abandon from the outset the demographic aspect. On the other hand,

[1] In 1910, miners accounted for 10·4% of the total engaged in Belgian Industry. *Recensement Général 1920*, p. 108.

by concentrating on the employees of one works and obtaining information on the travellers' industrial life from the firm's personnel records, such studies make up for the lack of comprehensiveness by the gain of thoroughness. The point of view of the factory inspector characterises two German monographs, one made before, the other soon after, the last war. A Dutch author deals with the matter in connection with a firm's housing-scheme for its employees. Inquiries into the daily movements of the miners in the South Wales Coalfield are included in this section, since they are based on information provided by the collieries, although in some respects they resemble demographic studies.

a Monograph on a Factory at Durlach, Baden, by F. Ritzmann [1]

The Gritzner factory, which provided detailed data for this investigation, manufactured sewing machines and bicycles at Durlach, a country town of 14,000 inhabitants near Karlsruhe. The firm's hourly-paid workers numbered 2,888 of whom 427 were females. Less than 40% of the workers lived at Durlach itself; about 20% were domiciled at places close-by; over 40% however, came from more or less distant places, walking up to 4·4 miles and travelling by rail up to 16·7 miles. Ritzmann undertook to examine the effects of the daily journey on absenteeism and on efficiency. For that purpose he classified the areas of residence of the Gritzner workers by mileage from Durlach and by remoteness from a railway-station. The data on days and hours worked by each employee were obtained from the personnel records of the firm and collated with the places of residence. During 100 working days, workers living at or near Durlach actually worked an average of 97·1 days; the inhabitants of the second- and third-best groups of places (in a combined order of length and discomfort of the journeys), 95·8 and 96·0 days respectively, the residents from distant places, only 94·8 days. Similar differences prevailed as to hours worked: during those 100 working days, workers of the four residential groups averaged 940, 912, 910 and 902 hours respectively. This means that long and tiresome journeys had a definite adverse influence on the attendance at work. The age-distribution, it is true, was not

[1] *Einkommens—und Wohnverhältnisse der Arbeiter der Maschinenfabrik Gritzner A.G. in Durlach*, 1914. The data refer to 1908.

homogeneous in the various groups of places ; but when age-groups were analysed separately, the same result emerged. An examination of wages earned at piece-rates showed a similar decrease in earnings with increasing distance and remoteness of the dwelling-place. Ritzmann looked into the question of whether the lower industrial efficiency of the inhabitants of remote villages was due to part-time farming, but could find no evidence of it.

The inquiry could not be carried out in such detail as to distinguish the types and grades of workers domiciled in the various places. But in Ritzmann's view this point was not responsible for the fact that those living at a distance showed inferior results as to attendance at work and hourly earnings. Nevertheless, he deprecated uncritical generalisation of his conclusions.

b Monograph on a Factory near Heidelberg, by Charlotte Grabe [1]

As in the Durlach study, the workers—about 2,000 men engaged in the manufacture of railway-carriages—were grouped according to the distance of their residential places from the workplace and to the strain of the journey ; thus places on the railway were distinguished from isolated villages dependent on cycling or walking (the neighbourhood is hilly).

The frequency of accidents and of sickness, the duration per case of sickness, and other industrial absenteeism grew with increasing length and strain of the journey. This became particularly clear from the examination of an apparent irregularity of the results. Workers living at moderate distance with fair travelling facilities showed an unexpectedly high liability to accidents and sickness, but only during the summer. The explanation was that in fair weather many workers living between three and eight miles away did not avail themselves of the railway but, in order to save fares, walked or cycled. This involved greater strain than a railway ride and so put those workers on the same level as those with the least favourable transport facilities.

[1] *Der Einfluss der Pendelwanderung auf die Arbeitnehmer*. Badische Wirtschaftsstudien, 3, 1926. The figures refer to 1923. See also E. G. Dresel and Charl. Grabe, "Einfluss der Pendelwanderungen auf die Arbeitnehmer" in *Deutsche Medizinische Wochenschrift*, 1924, No. 28.

The workers coming to work from Heidelberg had low accident and sickness rates ; this was attributed to the fact that these town-dwellers did, at the most, some gardening besides their job, whereas many villagers worked on their farms in the early morning, before starting for the factory, and were therefore already tired before industrial work began. The average duration of illness, too, was higher among persons living at a distance ; this need not mean more serious cases, but might result from a less strict control in remote places on the part of the insurance authorities or from the convalescents' fear of the additional strain caused by the journey.

Finally, absenteeism for reasons other than ill-health proved equally to have a rising tendency with the growing length and strain of travelling. The time lost in this way during the 100 days under observation was 3·3 hours per person for those living near the factory, and rose to 12·4 hours per head in the most remote places.

c Monograph on a Works at Eindhoven, Holland, by F. Bakker Schut

In investigating the daily journeys of the employees of the N.V.Gloeilampenfabrieken at Eindhoven, Holland, F. Bakker Schut [1] proceeds, on the whole, on the lines of F. Ritzmann and Charlotte Grabe. His findings corroborate their results : sickness frequency and sickness duration are higher among workers who have to travel to work than among those living at Eindhoven. The composition of these two groups of workers as to age, type and grade of work, and the possible influences therefrom, are not examined. What makes this study interesting is the attempt to express all adverse implications of the journey in terms of money.

In 1929, the works employed nearly 20,000 hands, more than could be recruited from Eindhoven, a town of some 80,000 inhabitants. F. Bakker Schut starts from the assumption that the firm, in order to secure a sufficient, suitable and stable labour supply, has got either to finance housing schemes in Eindhoven, or to subsidise transport services for its employees. His problem is, therefore, costs of the daily journey versus costs of house building. Consequently, he takes a wide view of "travelling expenses".

[1] *Industrie en Woningbouw*, 1933, pp. 81–120.

To fares he adds such "indirect costs" as are not incurred by those who live near the works: namely loss of time, greater fatigue and higher absenteeism, meals eaten away from home and increased labour turnover. For assessing the costs of these various items, the author has combed out the relevant continental literature and in addition collected unpublished Dutch material. Nevertheless, his procedure is not convincing. Knowledge of the facts and effects of daily travelling is, so far, scanty, and the few numerical findings do not seem safe for generalisation. In making up the "indirect costs" of the journey, he sometimes takes as a basis what ought to be an object of investigation. The costs of cycling, for instance, are put at the same figure as those of bus rides, on the argument that the bus fares are balanced by higher "indirect costs" of cycling.

As, however, Bakker Schut's computing of "indirect costs" of travelling to work has been the first, and so far the only, attempt of its kind, it is worth mentioning in this context. He adds to each employee's fares:

For loss of time: Fl. 0·05 per hour travelling;
,, meals consumed away from home: Fl. 0·31 per week;
,, increased labour turnover: Fl. 0·29 per week;
,, increased fatigue and higher sickness and accident rates: 2% of the wages.

The total "indirect costs" average Fl. 1.55 per week, which amounts to 60% of the average direct transport costs.

Although at present Bakker Schut's procedure of determining the "indirect costs" of travelling to work cannot be accepted, his method may attain some value in future when the circumstances of the daily journey are more accurately known.

d Inquiries into the Daily Movements in the South Wales Coalfield

i second industrial survey of south wales [1]

The journey to work is included in a detailed inquiry into *Labour Supply and Demand in the South Wales Coalfield.* The Monmouthshire and South Wales Coalowners' Association provided data on the journey to work of nearly 120,000 men and boys employed by them in July 1936. The study thus gives a picture

[1] Director: H. A. Marquand. 1037, Volume Three, Chapter I, by Hilda Jennings.

of the daily movements in the staple industry of the region. The information covers home address and workplace of the miners and also the mode of transport and the approximate length of the journeys in miles. While the chapter is mainly concerned with the separate analysis of conditions in the various areas, with a view to the potentialities of development of certain areas, there are also some results of more general import.

In every Employment Exchange Area within the Coalfield, a considerable proportion of miners live beyond walking distance of their work. The percentage ranges from 7·8 to 88·6 in various areas; on the average, 33·4% of the employed miners travel to work by public means of transport; in addition, some of those listed as living in vicinity of their work cycle.

The distances travelled vary greatly. There are some with daily journeys of over ten miles each way—43 men living in the Pontlottyn Exchange Area travel as far as 32 miles; but as a rule, the distance travelled is between one and ten miles, and the majority have journeys of considerably less than ten miles. On the whole, men living in the most heavily depressed area travel farthest. Through the closing down of collieries in their neighbourhood, some places have become mere dormitories for more distant collieries, and others are now actually outside the boundaries of the coalfield.

Most of the travellers use the bus; only 18·8% go by rail. This is a recent development. "Since road services became easily available and, in particular, since the advent of large numbers of privately owned miners' buses, which take them almost from door to door, the method of travelling has changed considerably. Whereas at one time the majority of men travelled by train, the great majority now travel by road" (loc. cit., p. 30).

The weekly cost of travelling varies from seven to eight shillings in a comparatively small number of cases to one shilling. The average train fare is 1*s*. 11·7*d*.; the average bus fare, 2*s*. 4·4*d*. and the total average for both road and rail transport 2*s*. 3·4*d*. per week.

In the authors' view, the very high degree of mobility in the Survey Area and the range of travelling suggest that journeys of up to ten miles each way would not be looked upon as prohibitive in length by prospective workers in new industries.

ii STUDY OF THE WESTERN PART OF THE SOUTH WALES COALFIELD, BY EMRYS J. HOWELL [1]

This survey covers the anthracite area, where 44 collieries employed, in 1938, some 23,000 colliery workmen. The home addresses of these miners, obtained from the collieries, show that "the geographical distribution of employees varies considerably even for neighbouring collieries. At some, a restricted mining community is nucleated within a few miles of the pit-head, while at others over 50% of the daily immigrants live over 11 miles away". The distance between home and workplace is given for 16,800 miners employed in 31 collieries. Nearly one-half of these live close to the colliery in which they work; another quarter live between one and three miles away; 1,090 or 6·4%, on the other hand, travel over 11 miles each morning and evening. No account is taken in this computation of distances travelled underground by workers at the coal face, which in the older collieries may be one or two miles.

In commenting on the redistribution of mining labour in the western part of the South Wales coalfield, E. J. Howell states that

> the tendency at the moment (i.e. before the outbreak of the present war) is for miners to prefer long journeys daily to finding homes near their new workplaces. This is due partly to lack of housing facilities and partly to the miner's natural aversion to moving his home, especially when the continually changing face of the coal-mining industry may render such a course inadvisable.

4 *A METHODOLOGICAL REPORT BY THE INTERNATIONAL STATISTICAL INSTITUTE, 1934.*

Besides the investigations just reviewed, other fact-finding inquiries have been carried out on similar lines, to mention only studies on the relevant conditions in Amsterdam, Copenhagen, Paris, Stockholm, and Zurich. An annotated bibliography is given in a *Rapport sur la Migration et le Mouvement Alternants* by a committee of the Institut International de Statistique.[2] This report is a study of the scope and methods of the investigations

[1] "Movement of Miners in the South Wales Coalfield" in *The Geographical Journal*, September 1939, pp. 228–237.

[2] XXIIe Session de l'Institut International de Statistique, Londres, 1934. Section 3, R A P P. Rapporteur: J. H. van Zanten.

made in various countries. It is mainly concerned with official statistics, but points to the desirability of complementing the data to be obtained from censuses by information from firms on their employees' daily journeys. Various aspects of the journey are mentioned as calling for examination, under three headings: economic aspects, including certain demographic features; social aspects; and transport. Comparatively little attention is given to the costs and hardships of the daily journey; the case of secondary earners and the problems raised by the emergence of "dormitories" are not discussed.

Chapter II

THE PRESENT STATISTICAL INVESTIGATION

PRIVATE collection of statistical information is bound to be restricted in scope and beset with difficulties. In the following pages, the results of inquiries made for this study in English industrial areas are presented and analysed.[1] Though not necessarily representative of the whole country, they are offered as samples testifying to the diversity and complexity of prevailing conditions.

1 *THE DAILY JOURNEY VIEWED AS CONFLUX AT THE WORKPLACE*

a Survey among Member Firms of the Industrial Welfare Society

To obtain an approximate picture of the conditions of the journey to work, a survey was made among some members of the Industrial Welfare Society. The answers of 36 firms which gave more or less useful information are arranged in a synopsis in the Statistical Tables. These firms, with a total of nearly 70,000 employees, represent various types of enterprise, including both factories and shops, located in large towns and in the country, in the North and in the South, and employing from 200 to 9,000 hands.

A few firms continue, as in the old days, to recruit the bulk of their workers from the village in which the factory is situated. Four out of twenty-five firms which give detailed figures have from 60 to over 70% of their employees living within walking distance of their work, and six others have 50, or nearly 50%. The extent to which bicycles are used is remarkable. In the case of two firms more than 60% cycle to work, and in eleven others from 30 to 50% More than once, however, attention is drawn

[1] For the tables and charts referred to in the text see p. 164 seq. and p. 200 seq.

to the fact that the figures of cyclists apply to the summer only and are appreciably lower for the winter months. Public transport services are used by the majoriy of employees in eight of the reporting firms, and by one-third to one-half in a further seven. One firm states that transport facilities are worse for night than for day workers.

Data about the duration of the journey are given by twenty-two firms. Twelve report that over 70% of their employees live at less than half an hour's distance, seven of them even over 90%. On the other hand, long daily journeys prevail in the large towns, especially in London, and in the open country where transport services are poor. Five firms report that 10% or more of their employees have a journey of at least an hour each way, the numbers of the workers concerned being 425, 410, 225, 160 and 125. In the London area, journeys of an hour and more are frequent.

The reasons given for the long duration of journeys are various: one firm, "in a country district . . . having grown during the last few years at a very rapid rate" thinks that "of course, a great many men . . . would live in the immediate vicinity of the works if accommodation was available". Other cases are juvenile or female workers who have found suitable employment at some distance from their homes but prefer the long daily journeys to living alone near their work. A third cause of long travelling is the remoteness of some of the villages from the nearest labour market as well as from transport services, so that any journey from such homes to industrial employment is long. A further instance is that of Trafford Park, the trading estate outside Manchester, which was built at some distance from workers' residential places. Lastly, there is the huge built-up area around London which causes many very long journeys within the conurbation.

Those who go home for lunch may be assumed to live within easy distance of their work; eight factories report that from 70 to over 90% of their employees do this, most of them walking or cycling. Although long and expensive journeys are a reason for remaining on or near the works premises during the mid-day break, this will depend also on the duration of the break and on the canteen and café provisions.

The main impression conveyed by this survey is the great variety in the journey to work among the employees of various firms and in various localities.

b Examination of single Firms

i CARRERAS LTD., TOBACCO AND CIGARETTE MANUFACTURERS, LONDON, N.W.1

At the time of the inquiry, Carreras had approximately 2,500 employees, the vast majority of whom were girls. The firm had moved to Hampstead Road, London, N.W.1, in 1929, seven years before the survey was made. Until then their premises had been in City Road, E.C.1.

Local Distribution of the Employees' Homes (Table 1 and Chart).[1]—The employees' homes are spread over the whole London area and beyond. But the distribution is uneven; as the following table shows the Eastern and Northern postal areas contain four-fifths of the girls' homes:

Number of Employees with Home Addresses in the following Postal Districts

Districts	Employees No.	Employees %	Districts	Employees No.	Employees %
East	867	33·5	North	789	30·5
E.C.	89	3·5	N.W.	223	8·6
S.E.	155	6·0			
	1,111	43·0		1,012	39·1
West	47	1·8	Outside London	305	11·9
W.C.	39	1·5			
S.W.	70	2·7			
	156	6·0	Total	2,584	100·0

More than two-fifths of the total employees live in the Eastern area, the majority in the Eastern districts proper. Nearly two-fifths live in the Northern and North-Western areas. 6% live in the Western, South-Western, and West Central areas, while the remaining 12% give addresses outside the London postal area: 143 girls in Middlesex and 8 in Herts; 71 in Essex and 49 in Kent; 34 in Surrey.

[1] p. 166 seq. and pp. 200/1.

When considering the large part played by the Eastern Area as a dormitory of Carreras' employees, one has to bear in mind that many have been with the firm for more than seven years: when taking on the job, they had to travel only from East London to City Road, E.C., the works' former site; when the factory was moved in 1929 to N.W.1, they had to put up with longer journeys. An interesting side-light is provided by comparing the residential distribution of new entrants with that of the total body of employees (see Table 2). Amongst those recruited during the year ending in October 1936, the Eastern area provided 34·2% as against 46·8% among the total employees in November 1935; North and North-West, on the other hand, accounted for 51·3% of new employees as against 38·3% of the others. This shows a marked, if not sweeping tendency towards drawing new labour from the districts adjacent to the factory. Certain inconsistencies between the three columns of the table are due to employees leaving the firm during the year under review.[1] A noticeable feature is the increase in the number of addresses outside the London postal area; their number, in 1936, surpasses the sum of the corresponding figures of the body of employees in 1935 plus new entries. The explanation is that some of the employees have, during the year under consideration, moved out of town, with their families; a fair number of the girls reporting came from Dagenham and Becontree.

The questionnaire asking for particulars of the daily journey to work was handed to all girl employees and adequately filled in by 834, i.e. approximately one-third of the total. The table on page 135 shows the residential distribution of these girls, in comparison with that of the total employed by the firm.

The table shows that the various postal districts are approximately, though not accurately, represented by the returned forms. As would be expected, those with long journeys are more interested to report on this subject than those who live within walking distance of the works. Thus, 50% of the completed questionnaires show addresses in the Eastern area as compared with 43% of the total on the firm's pay-roll. The grouping of the girls reporting by postal districts shows a fair homogeneity for each district as to the length of journey. Although slightly

[1] E.g. the fact that the higher percentage of new entries from the Northern district failed to increase the total percentage of North, from 1935 to 1936.

loaded towards overstating the number of long daily journeys, the 834 completed forms which are analysed in the following pages can, therefore, be taken as roughly representative of the travelling conditions of Carreras' employees.

HOME DISTRICTS OF 834 EMPLOYEES WHO COMPLETED THE QUESTIONNAIRE

Group of employees	Postal districts in which . . . employees live			
	East E.C. and S.E.	North and N.W.	West W.C. and S.W.	Outside London area
834 employees reporting				
No.	419	320	33	62
%	50·3	38·4	3·9	7·4
Total employees % . . .	43·0	39·1	6·0	11·9

Duration of the Journey (Table 3).—One-third of the girls who report leave home more than one hour before work starts ; 30 % allow for three-quarters to one hour, 20 % for one-half to three-quarters of an hour, and only 17 % for not more than half an hour.

These figures are derived from the answers to the question, " When do you leave home ? " One would expect them to be parallel with the answers to another question, namely, " How long does your whole journey take from door to door ? " However, there is a marked difference : the number of minutes needed for the journey is very often smaller than the time of leaving home would suggest ; 9·5 % say they need longer than one hour, as against 33·3 % belonging to this group on the time-of-leaving-home basis. 34·4% travel half an hour or less as compared with only 17·0% leaving home half an hour or less before the beginning of work. It is true that a large number state 7 a.m., the full hour, as the time they leave home [1] ; but they do not usually give the correspondingly easy answer " one hour " as the duration of their journey ; girls whose journey takes one hour, often say that they leave at 6.50 or 6.45. Similarly, where the full hour does not appear at all, the answers to the two questions are such as : need 55 to 60 minutes, leave

[1] In the subsequent tables, those girls who say they leave home at 7 a.m. are grouped together with those leaving between 7.1 and 7.14.

home at 6.55; need ¾ hour, leave at 7.5. There are several possible explanations for this discrepancy: trains, buses and trams may run so infrequently that girls have to catch one which makes them arrive at the factory ten or fifteen minutes too early; early trains may be chosen in order to benefit from the cheap workmen's tickets, or else to secure a more comfortable ride before the peak of the rush-hour traffic. Another reason is that some time is allowed, in addition to the normal or minimum duration, for possible delays, in order to avoid being late—drastic penalties for lateness were in force at the time of the inquiry. This circumstance suggests, and girls' remarks confirm, that the last reason is the most effective one, the others operating in the same direction.

Both measurements of the duration have been tried in examining the various aspects of the journey to work; the time of leaving home has throughout proved to yield more consistent results. In the following analysis of the questionnaires, the time of leaving home has, therefore, been adopted as the yardstick.

Means of Transport (Table 4).—The situation of Carreras' factory in a busy part of London, and the spread of the employees' homes over a large area, make for a wide range of methods of transport. All available means and combinations thereof are used, although cycling was reported by one girl only. For short rides, bus and tram are often given alternatively ("take whatever comes first").

6·6% of the girls walk from door to door, while 93·4% use public means of transport. The method used by the greatest number of girls (29·5%) is the train—either on suburban or L.P.T.B. lines; next comes the tram (21·2%), while the bus accounts for only 7·8%. In addition there are many who use more than one sort of conveyance (ignoring changes from one train to another or between two trams). 29·9% combine several conveyances on their journey to work, 20 girls using not less than four, that is to say, changing three times.

Among the groups formed according to the duration of the journey, the bus is mainly used for short rides; this can at least partly be ascribed to the fact that no workmen's tickets are issued on London buses—see page 32; the tram is most frequent in the next, and the train in the third and fourth groups;

a combination of several means of transport accounts for nearly half of all journeys which start before 7 a.m.

Changing of Vehicles (Table 5).—Changing of trains, trams and buses must be considered as one of the main hardships of the daily journey; it adds many minutes to the time needed for actual transit and it often doubles or trebles the inconveniences of hurrying, waiting, queueing, "fighting" to get in, and strap-hanging. Only one-third of the girls who use public means of transport have through-journeys, while two-thirds change trains, trams or buses once or several times, both in the mornings and evenings. This includes changes from one tram to another or one train to another as well as combinations of different means of transport. A number of forms show alternative statements, such as "if fine (or: if early) walk, if wet (late) ride", namely from Camden Town or Euston Square to the firm. In these cases, the conditions of fine weather and being early have been taken to prevail for the tables; the results, therefore, understate rather than overstate the number of changes.

As one might expect, the number of changes increases with the length of the journey, from 0·21 changes per girl on short rides to 1·48 on the longest. Among the 142 girls who start their journey at 7.30 a.m. or later, 51 walk from door to door; of those who ride, 80% have through-journeys; in the group leaving between 7.15 and 7.29 nearly half have to change, most of them once. In the following groups the proportion of through-journeys steadily decreases from over a quarter to less than one-tenth. The proportion of journeys with two or three changes increases. Of the girls leaving home before 6.45 a.m., 47·3% change two or three times on each journey.

To a certain extent, the frequency of changing is dependent on individual choice. The factory is situated about ten minutes' walk from two major traffic points (Camden Town and Euston). Some of the girls alighting there from their trains or trams walk the remaining distance to the works; others take a penny ride, thus adding one more change to the record of their journey (and a further penny to their fare), while probably arriving a few minutes earlier. Two girls living in the same house and using the same principal method of transport may, therefore, have different entries for the items: number of changes, length of journey, and fares. Other cases of changing are unavoidable, at least they were so at the time of the inquiry ; in the meantime, traffic improvements such as running of through-trains may have reduced the number of necessary changes.

The Walk from Home to Stop or Station (Table 6).—Another feature directly affecting the length of the journey is the walk from home to the stop or station where the bus, tram or train is boarded. 38·3% of all who use public means of transport have to walk from 5 to 9 minutes ; 28·7%, 10 to 14 minutes ; 18·6% have less than 5 minutes' walk, while 12·5% have walks of fifteen minutes or longer. Thus 41·2% of the total travellers, excluding those who walk from door to door, need ten minutes or more to reach stop or station. There is a steady increase in the proportion of long walks (ten minutes and over) with increasing length of the whole journey, from 12% of those starting at 7.30 or later, to 59% of those who start before 6.45 a.m. It is thus obvious that the distance of the home from the means of transport contributes considerably towards prolonging the daily journey to work.

Fares (Tables 7*a* and *b*).—93·4% of the girls who report pay fares. (In addition, some of those who walk in the mornings and evenings, pay fares for riding home for luncheon and back). Over one-third of all girls reporting pay less than 6*d.* per day, and nearly another third pay between 6*d.* and 8½*d.* ; 8*d.*, however, means four shillings a week, or 10% of a weekly wage of £2. Over a quarter of the girls have daily transport costs of 9*d.* or more, and among them 79 (9·4%) pay at least one shilling a day, eleven even 1*s.* 3*d.* or over. According to expectation, the incidence of higher fares is on those who start early—44% of the girls who leave home before 6.45 have fares of a shilling or more, 16% of those leaving between 6.45 and 6.59 a.m., and 4% of the 7.0 to 7.14 group. All these groups show further considerable percentages of girls who pay between 9*d.* and 11½*d.*, while nearly all those starting later pay lower fares.

However, the amount of fares is partly determined by the use or otherwise of workmen's tickets, and this depends on the means of transport rather than on the length of the journey ; train and tram issue workmen's tickets ; buses do not. Three-quarters of the fare-paying girls use workmen's tickets, either for the whole or for part of their daily journey. The bulk (85%) of those who do not benefit from such reduction pay less than 6*d.* per day, the remainder mostly 6*d.* to 8½*d.* Most of these girls leave home comparatively late, nearly half at 7.30 or later, and another 40% between 7.15 and 7.29. A third group state that they use workmen's tickets " sometimes ", " when lucky ", " if available ", " if I can board a train at 7.10 ". The explanation is that only a restricted number of trains and trams are available for holders of workmen's tickets, and they are often overcrowded. The difference in the amount of fares between workmen's and ordinary tickets is considerable ; girls state such instances as 2*d.* against 4*d.*, 4*d.* against 8*d.*, 7*d.* against 11*d.*, 11*d.* against 1*s.* 4*d.*

Ages of the Daily Travellers (Table 8).—36% of the employees are between 20 and 24 years old, 31% under twenty, 26% from 25 to 29 years ; ages of 30 or more are stated by only 5·5% of all the employees who report. The table shows that younger girls live nearer the workplace ; of those under 20, one-half leave home before 7.15 ; of the girls between 20 and 29 years, two-thirds, and of those of 30 and over, more than four-fifths.

The proportion of those starting on their journey before 7 a.m. rises from 27% of the under-twenty's to 39% of the over-thirty's. In interpreting these results account must be taken of the removal of the factory in 1929, and of the long connection of some employees with the firm. Many of the older employees who joined the firm before 1929 were drawn from East London, affording them fairly easy access to Carreras' former factory site in E.C. ; the present site is less accessible to them. Girls in their 'teens, on the other hand, taken on in recent years have been largely recruited from the neighbourhood of the firm's present site. The high incidence of long daily journeys on older employees as is found prevailing at Carreras must not, therefore, be generalised without further investigation.

Effect of the Daily Journey on the Workers' Health (Table 9).—The time spent in travelling obviously means a curtailment of the time available for sleep and leisure ; the amount of fares paid means a reduction of the net wage ; hurrying, waiting, changing vehicles, and strap-hanging are discomforts. It remains, however, to be established, whether and to what extent long and uncomfortable daily journeys have measurable effects on the health and efficiency of the workers concerned.

It had been hoped to contribute towards determining such possible effects by collating the data of the journey with the records of absenteeism and of wages earned at piece-rates which the firm keeps for each employee. Unfortunately, particulars of the latter could not be obtained and the use of the former has led to inconclusive results.

The absenteeism of 587 girls during three years (1933 to 1935) could be examined. This number excludes cases of tuberculosis which entail extremely long absence from work, and also absence caused by a stay in the firm's convalescent home, a holiday in that home being frequently offered after several years' service with the firm and not reflecting illness.[1]

The aggregate days of absenteeism were computed for the groups formed according to the length of the daily journey. The average annual absenteeism is 6·02 days amongst the girls leaving home each morning at 7.30 or later, and 6·92 days amongst those leaving before 7.0 a.m. Table 9 shows that, on the whole, girls with long journeys have a higher absenteeism rate than those with short journeys. But the rise is not great and, more important, it is not steady. The absenteeism rate drops from the group with the shortest journeys to the next one and again from the fourth to the last group. Moreover, further analysis proves that the available data do not provide sufficiently great numbers for a statistical investigation of this kind. The absenteeism of the 587 girls under review ranges from nil to 208 days during the three-year period. If the six girls are excluded who were absent from work for more than three months during one of the three calendar years—such prolonged illness is unlikely to be connected with the journey to work—a correlation between absenteeism and lengthy travelling can hardly be said to emerge from the table. A minor source of error is that some girls will have changed their address during the three years, so that their absenteeism in the earlier part of the period may not be actually related to the recorded length of their journey in 1936. Separate examination of the younger and the older employees reveals further irregularities.

It must therefore be stated that the material under review has proved inconclusive as to the influence of long, uncomfortable journeys on the workers' health. It is all the more desirable that

[1] I am indebted for assistance in this part of the inquiry, and also for handing out the questionnaire to the employees, to Dr. T. Garland who was then the firm's medical officer.

further investigation on the basis of more ample statistical material should be made in order to clear this important issue.

ii ACHILLE SERRE LTD., CLEANERS AND DYERS, WALTHAMSTOW, E.17

The data available from this firm do not allow an analysis of the same thoroughness as the Carreras material. On the other hand, there is some additional information.

The works of Achille Serre Ltd., like that of Carreras, was moved to new premises in another London district, in 1929. The new works at Walthamstow, E.17, is within 15 minutes' motor-ride from the old works at Hackney Wick, E.9. Achille Serre Ltd. employ several hundreds of men as well as of girls ; their works is located, not in a central district with heavy traffic, but on the fringe of the town with—at the time of the inquiry—not too good transport facilities. In January 1936, the firm had 782 employees ; 692 or 88·5 % filled in a questionnaire on "Travelling Facilities."

Table 1 [1] shows the proportion of the sexes among the employees : two-thirds of the total are female, of the clerical staff, even seven-eighths.

The types and grades of work are, on the whole, equal for both sexes, spotting, dyeing, and pressing being done by men as well as by women. Data on the ages of the employees and other particulars are unfortunately not available.

Residential Distribution (Table 2).—With the exception of 3·8 %, Achille Serres' Walthamstow employees live in the Eastern London area ; the postal districts E.9 and E.17, i.e. the districts comprising the former and the present premises respectively, account together for 56·1% of all employees reporting, 38·9% being in E.17. According to the management, the actual proportion is even higher, as most of the 90 employees not reporting may be assumed to live near the works.

The residential distribution shows some interesting differences between the sexes. E.9, the postal district in which the old premises of the firm were situated is the residence of 20·6 % of all male employees who report as against 15·6 % of the females. E.17, the new district, on the other hand, accounts for 31·9 % of the men's but for 42·0 % of the girls' addresses. It seems safe to assume that a number of married men who lived near the former

[1] p. 174 fol.

factory site, continue to live there and accept the daily journey to the present works premises in E.17. This is partly due to the housing shortage in Walthamstow.

But the inference from the figures that local recruiting of female labour is easier, needs some qualification, on account of the 90 employees who did not fill in the questionnaire: the managers of the firm are of the opinion that the majority of the employees not included in the return live locally and walk. The composition by sex of those who failed to make returns is, however, the reverse of the employees reporting: only 35 are girls, and 55 men. If the greater part of the 55 men would be added to the addresses in E.17, the share of men in this district would noticeably rise; the inclusion of most of the 35 girls on the other hand, would cause a much smaller proportional increase of the female employees living near the workplace.

On the whole, relatively fewer male employees live in the districts adjacent to the present works, and more outside the East London Postal District. Dagenham, for example, is the dormitory of 9 male as against only 4 female employees of Achille Serre's Walthamstow works.

Duration of the Journey (Table 3).—The most frequent duration is between half and one hour (44%); next come those with under half an hour (39%); 17% had journeys of one hour or more. Girls have not quite as many short journeys as men and correspondingly more long ones. Actually, the difference is considerably greater than shown in the table, because, as argued in the preceding paragraph, the inclusion of the non-reporting employees would increase the percentage of short journeys among men, while not materially changing the proportion among girls. The male employees are thus shown to have shorter journeys than their female colleagues. This is a rather unexpected result in view of the residential distribution of male and female employees by postal districts. The main explanation is to be found in the different means of transport used.

Means of Transport (Table 4).—As regards walking, the figures are according to expectation. To the higher percentage of girls who live in the vicinity of the factory corresponds a higher ratio of girls who walk from home to work, namely over one-fourth, as against less than one-sixth of the men.

In the use of bicycles, however, there is a pronounced difference between the sexes ; not less than two-fifths of all males who report cycle, while this method of transport is negligible among females. Cycling covers a given distance from door to door quicker than using public transport services (Table 5). This is attested by the inquiry in two ways : (*a*) some employees specify the duration of the journey by both methods of transport which they use alternatively, according to the weather ; in all these cases, cycling takes less time ; (*b*) various employees living near each other but using different means of transport state different lengths of their journeys : cyclists need less time than users of tram, bus or train. There are two possible reasons for the girls' general rejection of the bicycle for the daily journey, at the cost of waiting for public conveyances and paying fares : the fatigue of cycling and considerations of dress. Cycling, especially in bad weather, is not compatible with being smartly dressed. Probably both reasons operate together.

Conversely, public conveyances are used by 69 % of the female, and by 44% of the male employees. Moreover, girls often take a conveyance over a short section of the journey where a man will walk ; changing and waiting for the second vehicle may add several minutes to the duration of the journey. In any case, as far as the evidence of Achille Serre's goes, female workers take longer to reach the workplace than males, even though the distance from their homes, as the crow flies, be less.

The Walk from Home to Stop or Station (Table 6).—Of all employees who use public means of transport, 37·0 % say that they have to walk 10 minutes or over from home to stop or station, including 12·4 % who need 15 minutes or more. Of the girls, 41·8 % report long walks, as against 25·3 % of the men ; walks of 15 minutes or more are reported by 14·8 % of girls and 6·5 % of men. As there is no reason to believe that on the average male workers live at shorter distances from the means of transport than females, the different answers indicate that girls walk slower and thus take longer time to cover a given distance.

Going Home for lunch (Table 7).—156 or not quite one-fourth of the employees go home for lunch. As would be expected, all of them belong to the groups with less than 30 minutes travelling time, more than half with less than 15 minutes. Of the latter group 88 % have their lunch at home, as against 41 % of the group needing between 15 and 30 minutes ; the number of girls going home is disproportionately high ; of the girls with journeys under half an hour, 70 % lunch at home, of the men, only 34 %. As an explanation it may be suggested that entries of, say, 20 minutes bicycling by male employees may stand for journeys of such strain as to make it not worth while to repeat the return journey for lunch ; further, streets are probably more crowded at mid-day so that the journey would take up more time than in the morning.

iii STANDARD TELEPHONES AND CABLES LTD., NEW SOUTHGATE, LONDON, N.11

At the time of the enquiry (January 1936) the Southgate Works of the firm had 4,124 employees of whom 966 or nearly one-quarter belonged to the staff. Males accounted for over two-thirds of the staff, but constituted only a moderate majority of the hourly paid workers, resulting in an average of 58 % of the total employees (Table 1).[1]

Until 1934 the firm had one factory at New Southgate (established in 1923) and another at Hendon, Middlesex, with about 1,500 workers at each. In 1934, the Hendon factory was moved to New Southgate, so that now both plants are combined there.

As only part of the employees under review (those formerly working at the Hendon Works) were affected by the removal of the factory to New Southgate, a smaller proportion of the total body of employees are found to live near the firm's former premises than is the case at Carreras' and Achille Serre's ; this is so in spite of the more recent removal of Standard Telephones.

The distribution of the operative workers' homes according to distance from the factory was the following :

Up to 1 mile	8·3 %
1 to 2 "	11·7 %
2 to 3 "	18·8 %
3 to 4 "	18·4 %
4 to 5 "	15·4 %
Over 5 miles	27·4 %
	100·0

Residential Distribution by Postal District (Table 2 and Chart).—The residential distribution by postal districts—covering workers

[1] p. 178 f. and pp. 202/3.

only—is shown separately for males and females.[1] It appears from the chart that local recruiting applies more to girls than to men: females preponderate among employees living near the works and are comparatively few in distant districts. The main reason is that girls are engaged in less skilled occupations and can, therefore, be found in any residential area, while the male employees of the light engineering industries include a high proportion of specialised and skilled men who have to be drawn from a wider labour market.

This argument is supplemented by another as far as the districts near the former site of the works in Hendon are concerned: a number of men had bought houses in the neighbourhood of the old workplace, either outright or on the hire-purchase system. After the removal of the factory to Southgate, they could not or would not leave their houses but took to the longish daily journeys from Hendon to New Southgate and back. This applies especially, according to the managing director, to members of the staff; but a glance at the chart which represents only the homes of the employees paid by the hour, shows it to be true of workers too.

Means of Transport and Duration of the Journey (Table 3).—After the concentration of their labour force at New Southgate, Standard Telephones had considerable difficulty in getting adequate transport services for their employees. At the time of the inquiry the firm stated that roughly 15% walked to work, and nearly as many cycled, leaving about 72% as users of public means of transport, the majority of whom came by train.

60% had journeys of half to under one hour, 30% less than half an hour, and 10%, namely 410 employees, one hour and over.

iv AUSTIN MOTOR CO. LTD., LONGBRIDGE, BIRMINGHAM

This inquiry is based on a traffic census taken by the firm in June 1937. The works have been at Longbridge, about seven miles from the centre of Birmingham, since the firm's foundation in 1905. At the time of the inquiry it had 20,000 employees, the great majority men. The special purpose of the census, to be discussed later, determined the range of the informaton

[1] These data, obtained three months later than those on the journey, refer to a slightly different number and composition of the workers.

obtained. The firm was not interested in the duration of the journeys ; but records are exact of the places where Longbridge employees live and—with a notable qualification—of the various methods of transport.

Places Involved (Table 1 and Chart).[1]—About 100 localities [2] contribute to feeding the daily stream of 20,000 employees towards Longbridge. Places send from one to over 3,000 inhabitants to work at Austins'. As the following table shows, more than half of the places send under 100 persons each, but contribute altogether only one-twelfth of the total Austin employees ; on the other hand, three places, where more than 1,000 Longbridge employees live account for nearly one-third of the total. The majority (59 %) come in groups of 100 to 1,000, fairly evenly in numbers of 100 to 300, 300 to 500, and 500 to 1,000.

Longbridge Employees by size of Groups coming from Individual Dwelling places

Number of daily travellers	Residential places		Employees	
	Number	Percentage	Number	Percentage
Under 20	25	25·8	134	0·7
20 to under 50 . .	19 } 57	19·6 } 58·8	573 } 1,620	2·9 } 8·4
50 " " 100 . .	13	13·4	913	4·8
100 " " 300 . .	20	20·6	3,448	17·6
300 " " 500 . .	11	11·3	4,041	20·6 } 59·0
500 " " 1,000 . .	6	6·2	4,074	20·8
1,000 and more . .	3	3·1	6,368	32·6
Total . . .	97	100·0	19,551	100·0

Although within the city boundary, the works is not immediately surrounded by large residential quarters—Longbridge Estate accommodates but 210 Austin employees, and Rubery, one mile away, under 1,000. However, the most substantial clusters are in fairly near-by places : Northfield and Westheath, grouped together in the firm's list and about two miles from Longbridge, house 3,227 Austin employees ; Sellyoak,

[1] p. 181 f. and pp. 204/5.

[2] In the firm's list, some districts of Birmingham are counted as separate units, while in some other cases several districts or localities are lumped together.

Bournville, and Bournbrook (4 miles) 1,310; Bromsgrove, however, with 1,831 inhabitants working at Longbridge, is six miles away.

A grouping of the places of residence by mileage from Longbridge results in the following distribution:

Unit	Employees living at ... miles' distance from Longbridge					
	3 and less	4–5	6–7	8–10	Over 10	Total
Number .	5,784	3,830	5,772	2,989	1,176	19,551
Per cent .	29·6	19·6	29·5	15·3	6·0	100
	49·2		44·8			

30% of the total live within a three-mile zone from the workplace and altogether 50% within a five-mile zone; nearly 45% live between six and ten miles away; 6% or nearly 1,200 employees live over ten miles from Longbridge, including 153 who live at distances of twenty miles or more. This measurement, as the crow flies, gives a rough idea of the journeys involved; no account is taken of the distances between the individual travellers' homes and their nearest tram, bus or railway stops.

As Longbridge lies to the south-west of Birmingham, numerous contingents of the labour force are found to live in the suburbs on the south-west fringe of the town. But the firm also draws largely on the labour resources of Birmingham itself; Austin workers' homes are spread over the whole built-up area, including the north-eastern parts. Another important area of supply of workers is the Black Country, to the west of Birmingham and to the north of Longbridge (Dudley, Old Hill, Cradley Heath, Stourbridge, Blackheath, etc.). As many as 500 come even farther, from Wolverhampton and surrounding villages. The reasons for the employment at Longbridge of workers from the Black Country, according to the personnel manager of the firm, are the following: people from that district, traditionally accustomed to foundry and core making, are better able to do this hard work than townsfolk, and they can earn higher wages at Austin's than they would at their places of residence. The rural areas to the east and west of Longbridge do not contribute labour to the Austin Works, and comparatively few come from the southern direction.

Persons Involved.—The big majority of Austin workers are men and boys; the (clerical) staff includes many girls. The firm's traffic census gives separate data for the workers and the clerical staff, the former split up into day and night shifts; these distinctions reveal noticeable differences in the journey to work of various types of employees.

Of the total 20,000 employees, the clerical staff numbers 1,310 (6·7%). Their working hours are later than those of the workers, from 9 a.m. to 5.30 p.m. Of the 18,241 workers, 84·5% work during the day; the night shift amounts to 2,296 (12·6% of the workers); a small remainder, 511, are engaged in three continuous shifts covering the 24 hours. The various groups show noticeable differences in their residential distribution. Of the 97 places (in the firm's manner of listing) where Longbridge employees live, 83 contribute the 2,296 night-shift workers, while 67 places are inhabited by the 1,310 members of the staff. The following table gives the mileage between the residential places and Longbridge, distinguishing the three groups of employees.

Type of employees	Total numbers	Employees living .... miles from Longbridge				
		3 and less	4–	6–	8–10	over 10
		Percentages				
Workers:						
Day Shift . . .	15,945	28·4	20·0	30·7	15·2	5·7
Night Shift . .	2,273*	23·0	16·7	29·3	21·3	9·7
Staff	1,298*	55·8	20·4	14·6	6·2	3·0

* Excluding 23 and 12 unascertained cases, respectively.

The staff's conditions as to the distance between home and workplace are favourable: 56% live not more than three miles away, as against half this rate amongst the day workers; only 3% of the staff travel more than ten miles, of the day workers nearly 6%. The night-shift workers, on the contrary, live at greater distances than the others: nearly 10% travel ten miles or more, and 21% between eight and ten miles, as against 15% of the day workers and 6% of the staff. The chart illustrates these proportions. The total staff numbers not much more than half (57 %) of the night-shift workers ; but in the places adjacent to Longbridge, the spots representing the staff are even larger than those which represent night-workers, while in distant places the spots representing staff members are unproportionately small, or missing altogether. Few members of the staff live in the City of Birmingham. Many of the night-shift workers, on the contrary, do ; this is probably due to good transport facilities in the big town even outside the rush-hours. The Black Country, too, sends quite a number of night-shift workers : as work performed on night shift is of the same kind as that performed by day, core makers and foundry workers largely drawn from these places are needed also for night work.[1]

Means of Transport (Table 2).—Just over 60 % of the employees use public transport services [2] for their journey, nearly 30 % come by private vehicles (motor-cars or bicycles), and 10 % walk from door to door. At face value, the most frequent method is the tram (27 %), with bus or coach not much behind (25 %), while the railway seems to account for only 8 %. This result, however, is loaded against the last-named method, by the terms of the census form. The purpose of the census was to investigate traffic congestion near the works ; the form issued to the workers contains, therefore, the instruction : "If you use two services place a cross against the service by which you arrive at these Works." Consequently, travellers who go by train from their residential place to Birmingham and there catch a tram for Longbridge are recorded as using the tram, while in fact the tram is only their secondary means of transport ; a case in point is a man recorded as coming the 21 miles from Warwick by tram. The bus must be assumed to do similar tributary service and, therefore, the figures for both tram and bus are overstatements, that for the railway an understatement. According to the personnel manager, however, not many journeys involve changes. It is certainly an advantage of bus and coach that they can pick up travellers near their homes and carry them straight to the

[1] Some night-workers remain permanently on night duty, while others get transferred to the day shift at times. Perhaps the differences of local distribution would be even more marked if those workers could be eliminated who change from night to day work occasionally. The listing, however, includes all who actually worked on night-shift at the date of the census.

[2] Including coaches run by private operators ; see next paragraph.

works. Austin employees were served, at the time of the census, by not less than 66 private coach operators. Most of this transport conveys Austin employees only and is not available for the general public.

The variety of means of transport used for the daily journey from individual places is remarkable. Often the most frequent method from a given domicile accounts for less than half of the journeys from that place to Longbridge. This is true not only of districts of Birmingham where one would expect varied transport services, but also of smaller places; of the 312 Austin employees who live at Blackheath, for example, 106 come by bus or coach, 87 by train, 62 ride by car or motor cycle, and 57 use a pedal bicycle. The presentation of the figures does not make it possible to determine how far the choice of the method of transport is influenced by such factors as the distance from the railway station of the respective homes, and the amount of wages earned. Differences between workers and staff in this respect will be discussed presently.

Private means of transport are used for the daily journey by 29·2% of the total employees: 13·4% go by pedal cycle, 2·0% by motor cycle, and 13·8% by private car. The proportion of car-users is uncommonly high for this country and would probably not be found prevailing except among employees of a works manufacturing motor cars. Over 2,000 of the day-shift workers report private cars as their means of transport, but the number of cars parked in the morning is only about 500, according to the personnel manager. The explanation is that several employees living at the same place or on the same route share one car, probably often sharing the costs. This procedure was an offence in peace-time, unless the Public Service Vehicle Licence and the Road Fund Licence were paid for. The private bus and coach operators oppose the sharing of cars, but usually little can be done, as the persons concerned generally affirm that the driver just gives a lift to his friends. The low proportion, of those who walk from door to door (10·2%) reflects the fact that Longbridge is not in a densely populated district.

Workers on day shift and on night shift, and the staff avail themselves of the various means of transport in different proportions. Public conveyances are used by 60% of the day workers, by 64% of the night shift, and by 52% of the staff.

While one-tenth of the day-shift come by train, the rate is only 4.4 % among the staff, and next to nothing among the night shift. The tram is more important for the night shift and the staff (over one-third in both cases) than for the day workers (one-fourth). The column " bus or coach " shows rates of 26 % and 28 % for day and night-shift workers respectively, but only 10 % for the staff. When seeking an explanation, one must remember that the staff starts work an hour later than the workers ; on some routes it would not pay to run special coaches for the comparatively few members of the staff inhabiting the various places.

18 % of the staff live within walking distance ; this is about double the ratio of workers in both shifts. Cycling on the other hand, is much less frequent among the staff than among workers. A high proportion of the staff come by car, namely 21·1 %, as against 13.6 % of the day-workers and 11·4 % of the night-shift ; if the fact is taken into consideration that many of the clerical staff are girls, few of whom probably come by car, the rate of car-users among the male staff must be put considerably higher than the 21 % quoted above. The staff thus enjoys a double advantage as to the daily journey to work : a high proportion live near the workplace, and many arrive at the works by private car. The opposite applies to the night-shift : proportionally more of them live at distant places than of the day-shift workers and fewer enjoy transport by motor-car.

Further Features of the Journey to Work at Longbridge.—Not only the results of the census are interesting, but also its origin and the practical use made of it. The simultaneous arrival of so vast a number of persons produced considerable traffic difficulties. When, for armament purposes, the plant was to be extended, it turned out to be impossible to increase the transport services at the peak hours. The expedient adopted for solving this problem was a " staggering " of the working hours.[1] The Longbridge works consisted of three plants with hitherto uniform working hours (for the day shift) from 8 a.m. to 5.30 p.m. Now plants A, B, and C were to start work at different times, and the transport services had to be adapted. The traffic census was taken in order to find out where the employees of each of the

[1] See : G. Shrosbree " Staggering of Working Hours ", in *Industrial Welfare*, November 1937, pp. 22–25.

three plants lived, and how they travelled. Thereupon the transport services were re-arranged to suit the respective starting times under the new plan. At first, three groups were formed, one working from 7.30 to 5.0, a second from 7.45 to 5.15, and a third (and largest) as before, from 8 to 5.30.[1] On the whole, the scheme has been successful, but an interesting alteration became necessary; the second relay had to be given up. It did not pay bus-operators serving small places, to run special buses for the 7.45 people; these had, therefore, to come with the 7.30 shift, but had to wait in the evening for those who finished at 5.30, and thus lost half an hour every day by waiting. Consequently, the 7.45 start was cancelled and this group begins work at 7.30.

Evidence on some other aspects of the journey to work was obtained from the personnel manager of Longbridge. The labour turnover is 35% per year including those laid up during seasonal slack times, and re-engaged later on. These workers enter other employment in the meantime and need, therefore, access to alternative labour markets. A middle-aged man may have another reason for not moving nearer to his workplace, namely consideration for his earning children: he has lost his work at the old place, while they have retained theirs—he therefore undertakes the daily journey to Longbridge.

Many workers have only a snack before leaving home and take their main breakfast at the factory before starting work, thus using the minutes of waiting. Employees who go home for lunch mostly live within walking distance, but some go by bus or tram, and a few cycle. Before the war, the majority of workers brought their lunch with them, usually remnants from the preceding night's supper. Against taking lunch in the firm's canteen, even at reduced prices, was the custom that the housewife was expected to provide all meals out of the money handed to her by the earners; to pay for lunch in the canteen, was, therefore, a drain on the worker's pocket-money.

The journey from door to door is actually not the whole journey for employees at Longbridge; since the factory has a road frontage of over a mile and covers 100 acres of a 200-acre site, anything up to 10 minutes' walk is necessary from the gates

[1] Workers, while agreeing to start half an hour earlier, objected to staying at work after 5.30 p.m.

to the proper workplace. This walk has to be added to the duration of the journey, for clocking-in is done at the departments. Up to two minutes' lateness is ignored, workers who clock in between 8.3 and 8.15 have a quarter of an hour's wages deducted, and so on.

V DUNLOP RUBBER CO., LTD., ERDINGTON, BIRMINGHAM

Transport grievances of another kind caused the Dunlop Rubber Company to inquire into the home addresses of the several thousand employees of Fort Dunlop which is situated in the north-east of Birmingham. In a certain department with female workers, two shifts were worked (in 1935), from 6 a.m. to 2 p.m. and from 2 p.m. to 10 p.m. When the Home Office made an inspection before approving the new hours, girls complained about inadequate transport services at the unusual hours of beginning and ending work. An inquiry ensued into the residential distribution of the employees, distinguishing between the sexes and the various departments ; the results were entered on maps of Birmingham, but not tabulated. The charts prepared by the firm bring out, however, one feature of considerable interest : specialised workers who cannot expect to find similar work with other firms in the town tend to live near Fort Dunlop, while the homes of those with more general occupations are spread over a much wider area.

vi MORRIS MOTORS LTD., COWLEY, OXFORD [1]

The Morris Cowley works are three miles south-east of the centre of Oxford. At the time of the inquiry, the works had 5,400 employees, including a staff of 570, on the pay-roll. While of the staff, one-quarter are girls, the proportion of females among the workers is negligible, under 2 %. Their percentage among the total employees is 4·3.

The distribution of workers' homes by mileage is as follows :

[1] The available statistical material makes it possible to deal either with the employees of the Morris Cowley works only, or else with the aggregate employees (about 10,000) of three factories, namely the Morris works at Cowley, the Pressed Steel works at Cowley, and the Morris Radiator Branch, Woodstock Road. The latter alternative was adopted in the *Survey of the Social Services in the Oxford District*, Vol. I, p. 309 *seq*. The present study, however, is limited to the Morris Cowley works. Otherwise the results for Cowley, the south-east corner of Oxford, would have been distorted, because the Radiator Branch is in North Oxford and therefore a separate unit from the point of view of the journey to work. See table, p. 184 f.

81 % live within a five-mile radius of the works, 7 % between 5 and 20 miles, and 12 % over 20 miles away. As compared with the labour of the Austin Motor Co., at Longbridge, Birmingham, a much higher proportion live fairly near the Cowley Works, on the one hand, but on the other, considerably more come from great distances. Both features are probably due to the smaller size of Oxford : the local employees live closer by, but in addition, recourse must be had to more distant dwelling-places.

The list of the residential distribution refers to October 1935, when the number of employees was 5,586, as against a figure of 5,400 in March 1936 on which the above percentages are based. 1,309 or 23·4 % live at Cowley, in the immediate neighbourhood of the works ; another 785 at Headington and Iffley, at distances of not more than two miles from Cowley ; a further 2,000 come from other parts of Oxford, so that on the whole over 4,000 or 73 % of the works' employees are domiciled in the town. The remaining quarter are drawn from 125 smaller places which contribute from one to 177 (Abingdon) employees to the Morris Cowley Works. The distribution of workers' homes within Oxford gravitates definitely towards the south-east.

No details are available on the means of transport or on the incidence of long journeys on various types of earners. But light can be thrown on other circumstances closely concerning our problem. The fluctuation of employment is illustrated by the following figures :

Number of workers employed on 1st January, 1935	5,105
„ „ „ „ „ 31st December, 1935	5,951
Increase during 1935	846
„ „ „ engaged during 1935	2,324
„ „ „ who left the firm during 1935	1,478
Total changes	3,802

These are interesting figures : while the number of employees increased during the year by 846, this was concomitant with 3,802 changes of employment, 2,324 new entrants plus 1,478 leaving the firm. The number of changes thus amounted to three-quarters of the total on the pay-roll at the beginning of the year, and to 4·5 times the actual increase of employees. Such an instance of high labour-turnover should be kept in mind in face of the demand, so often heard to-day, for close proximity of residence and work. Some employees stay with the firm continuously for years and benefit from being housed nearby; but there are numbers who are laid-off and re-engaged according to seasonal and other business fluctuations; and there are yet others who do not remain long in any job. For them, accommodation with easy access to more than one workplace is required.

2 *THE DAILY JOURNEY VIEWED AS DISPERSION FROM THE PLACE OF RESIDENCE*

The conflux at the workplace is but half the picture: the complementary aspect of the daily journey is the dispersion from the place of residence. The former approach envisages the problems relevant to the labour market, the latter, the implications for private life. While, however, the collection of data at the workplace is comparatively easy, it is difficult to obtain statistical information at the widely scattered homes. Only a Census can provide a comprehensive survey of the daily movements from and to a residential area; as stated above, no relevant information has been collected in this country since 1921. The radiation of daily journeys from individual homes has not so far been the subject of investigation.

Full documentation being unobtainable, recourse must be had to such evidence, however fragmentary and out of date, as will give at least an idea of the incidence of the daily journey on the families living in various places and on the various members of the family.

a Sample Inquiry based on the New Survey of London Life and Labour, 1929–30

Questions as to the journey to work were included in the household investigation card of the New London Survey. In the presentation and description of the results, however, these items were discarded. The following analysis is based on the original cards of the four boroughs East Ham, Stepney, Tottenham and Willesden. The Survey was made by sample method, covering about one in thirty families. The statistical material is limited in several respects. The data refer to 1929–30, over a decade ago; the survey is restricted to working-class families, the classification being based on the occupation of the father; this means that the families include children engaged in non-manual occupations. It is, therefore, possible that certain differences in the journey to work found between the groups of fathers and of sons and daughters are due to this bias of the statistical material.

In view of these and some other shortcomings of the material, the following analysis has been restricted to one aspect of the problem on which the more recent data obtained from firms and their employees fail to throw light, namely the incidence of the daily journey on the various earning members of the family. For this purpose, only families with two or more earners need be examined. The weekly amount of fares has been taken as the measuring rod of the journeys, although the amount of fares paid is considerably influenced by the use or otherwise of workmen's trains.[1]

In spite of so many qualifications the following analysis is offered as a first statistical attempt to shed light on an important aspect of the journey to work.

FAMILIES' AGGREGATE FARES IN THE FOUR BOROUGHS (TABLE I)[2]

The sample inquiry in the four boroughs of Stepney, East Ham, Tottenham, and Willesden yields information on 923 families with more than one earner each.[3] The average weekly expenditure for the transport to work is 4*s.* 2*d.*, including the families who do not pay fares. As an item in the family budget the amount of 4*s.* 2*d.* is by no means inconsiderable—the maximum acceptable amount of rent for the poorer family is usually put at 10*s.* per week. The sample inquiry does not indicate

[1] The composition of the users of workmen's trains by sex, age and occupation is not known. Nor is it possible to assess the weight of various conflicting influences on the use of workmen's trains by sons and daughters of the sample families. Some are in non-manual employment which means that they start work only at 9 a.m., while all fathers are wage-earners. Workmen's trains, however, cease at 8 a.m. (see p. 61). The use of workmen's trains is, therefore, less attractive to "children" who are clerks or shop assistants than to their fathers, although the lower wages earned by juveniles force numbers of them to travel by workmen's trains in spite of the waiting involved. On the other hand, juvenile workers are entitled to reduced fares (by one-half) on season tickets for one month or longer.

[2] p. 186 f.

[3] The numbers of families included are not weighed according to the total population of the respective boroughs. The number of earners per family which influences, of course, the aggregate amount of fares, is smallest in East Ham (2·65) and highest in Stepney (2·90).

that fares tend to rise with increasing incomes. Less than one-quarter of the families have no fares, with all earners walking (or cycling) to their work. Another two-fifths of the families pay aggregate fares of 4*s*. or less, while 37 % have transport costs of over 4*s*.

The four boroughs show marked differences, especially Stepney and East Ham. In Stepney, over one-third of the families have all earners working in the neighbourhood of the homes, and the majority of fare-paying families spend not more than four shillings, so that little more than one-fifth of all families pay over 4*s*. a week. The average weekly fares per family are 2*s*. 8*d*. These proportions reflect Stepney's character as an old-established part of London with many opportunities for work within the district and, moreover, adjoining the City of London. East Ham, by contrast, is a borough of comparatively recent growth on the outskirts of the built-up area. The differences, as compared with Stepney, in the age of the district, its location, and its social and economic structure are responsible for rather different conditions as to the journey to work. Apart from transport workers employed in the docks, "East Ham is largely a dormitory for clerks, shop assistants, mechanics and others whose employment lies elsewhere." [1]

East Ham.—The sample inquiry in East Ham includes 195 families with more than one earner (unemployed persons and lodgers excluded) ; 177 of these have the father and children or other relatives working, while in 18 families the father is dead, incapacitated, unemployed, or at sea ; the working members in these families are wives, children, or other relatives. The earners of the 195 families consist of:

	177	male householders
	154	sons [2]
	180	daughters [2]
	5	wives and widows
Total	516	persons

Since unemployment among sons and daughters is fairly equal (5 and 4 respectively), the preponderance of the number

[1] *New London Survey*, Vol. III, p. 393.
[2] Including other male or female relatives, respectively.

of daughters reflects a tendency among grown-up sons to leave their parents' house and live by themselves—probably nearer their work—while unmarried daughters continue to live with their parents. The average cost of transport per family in the East Ham sample is 6*s*. 8*d*. a week ; only in 8·2 % of the families all earners work locally, and less than one-third have aggregate fares of 4*s*. or less, more than two-fifths pay over 4*s*. up to 10*s*., and nearly one-fifth over 10*s*.

The distances between home and workplace of the various members (expressed by the amount of fares) are shown in Table 2. Of the total 516 earners, not quite 30 % live locally or cycle. Of those who use public means of transport, the largest number (37·2 %) spend over 2*s*. 6*d*. to 4*s*. a week. About 31 % pay 2*s*. 6*d*. or less, and another 31 % more than 4*s*. Among the three groups of earners, the sons show about the same proportions as the total earners ; of the householders, more live near their work, namely, 37·8 % ; of the daughters, on the other hand, only 22·2 %. But the conclusion that the girls' workplaces are farther away from their homes needs some qualification ; only one girl cycles, whilst seven fathers and ten sons do so, thus increasing the group "no fares", while actually the distance may be considerable. Similar is the case of railway employees (one father, three sons, no daughters) who benefit from free passes. Further, female earners are conspicuous in the groups of low fares. This fact suggests that girls often ride short distances which men and boys will cover by walking. The per-head expenses for the daily journey (Table 3) are equal for fathers and grown-up children of both sexes (about 3*s*. 9*d*.) ; juveniles pay less : girls only 2*s*. 9*d*., boys 3*s*. 4*d*. As the earnings of females are on the whole considerably lower than those of males, the same amount of fares means a greater financial burden for girls.

A considerable variety prevails in the incidence of the daily journey on the various members of the family. In some families the father works within walking distance of home, while the children have high fares ; sometimes the opposite is the case. In other instances, high fares are spent by the son, obviously in order to have a good training or a good job, while the daughter accepts any work in the neighbourhood ; or else the son may find work with his father's firm near home, whilst the girl has to travel.

Two of one family working at the same firm strengthens, of course, the case for living near the workplace (Table 4). Six sons and four daughters are employed by their fathers' firms ; in 19 cases, brothers, sisters, or other relations have the same employer. Twelve householders work on their own premises, as shopkeepers, shoe-repairers, bakers or the like : so do one wife and one daughter, and one son of a non-earning father. Eighty-six of the earners are recognisable as juveniles, 40 boys and 46 girls ; the proportion of those working locally is about the same as among adults.

Stepney.—The sample includes 392 families with two or more members gainfully employed ; in 332 of them, the father and at least one other member are earning, while in 60 families no male householder but several children or other relatives earn incomes ; comparatively many housewives are gainfully employed in the Stepney sample. The earners consist of :

	332	male householders
	384	sons [1]
	368	daughters [1]
	52	wives and widows
Total	1,136	earners

A considerable number work at home :

Householders	48 or	14·5 %	of all	male householders
Wives and widows	8 ,,	15·4 %	,, ,,	wives and widows
Sons	22 ,,	5·7 %	,, ,,	sons
Daughters	8 ,,	2·2 %	,, ,,	daughters
	86 or	7·6 %	of	the total earners under review.

14·5 % of the male householders in the Stepney sample give as their workplace, " at home ", " on his own ", or " self ", the latter being stated by hawkers and stall-holders in the neighbouring markets. Wives and widows even more frequently have their jobs at home, as shop-keepers or dressmakers. The proportions of sons and daughters working at home are much lower ; sons have

[1] See note (2) on p. 157.

more opportunities than daughters of joining their father's business.

Including the above-mentioned numbers of those who work at home or as hawkers, well over one-half (57 %) of all persons under review work locally, and of the male householders even two-thirds, as against 29 % and 38 % respectively in the East Ham sample (Table 2). Moreover, the amount of fares paid is only 2*s.* 2*d.* per fare-payer in Stepney, as against 3*s.* 7½*d.* in East Ham (Table 3). Just as in East Ham, householders in Stepney work within walking distance more often than do their children ; those householders who use public means of transport pay 2*s.* 4*d.* per head per week, adult sons and daughters spend 2*s.* 3*d.* and 2*s.* 2*d.* respectively. The fares of juveniles average 1*s.* 10*d.* Of the earning housewives, married or widowed, 57 % work either at home or within walking distance from home ; this proportion is lower than that of the male householders but higher than that of the children. The amount of fares spent by married women for their journeys to work ranges with those of juveniles (under 2*s.* per week) ; with their double range of duties, housewives who take a job find it specially necessary to avoid long journeys.

Instances of two members of a family working together (Table 4) are rather frequent : 21 sons and 7 daughters help in their fathers' business—some tailors and cabinet-makers have several of their children working with them : 22 sons and 5 daughters have the same employer as their father. Remarkably frequent are the cases where sisters work in the same firm (in two of the 26 cases, even three sisters have the same workplace). This happens much less often with brothers.

Although the entries of ages are rather scanty, many juveniles are recognisable as such by the description of their occupation (e.g. " shopboy " or " apprentice "), by the age of their fathers or by the low amount of their wages. Many juveniles work locally, 56 % of the boys and 59 % of the girls, while the proportions of adult sons and daughters are 52 %.

Tottenham.—169 families lend themselves to examination. In the case of 140, the father and children or other relatives are gainfully employed ; in 29 families, the male householder is dead, incapacitated or unemployed, but two or more other members of the family earn. The total earners number 471 :

140 male householders
157 sons [1]
159 daughters [1]
15 wives and widows

The proportion of those who work locally (47·8%) is less than that in Stepney but much above that in East Ham. It is remarkable that in the Tottenham sample householders work less frequently near their homes than their children, namely in 39% of the cases as against 50% and 54% of the sons and daughters respectively. The explanation may be found in the fact that within easy reach of Tottenham new industries have sprung up which offer employment to the younger generation while the fathers continue to work in other parts of London. A number of girls, however, work farther away than their fathers and brothers: one-fourth of the travelling girls pay over 4*s.*, their workplaces being in the West End. The number of persons working on their own is negligible. Cases of two or more members of a family working in the same firm are infrequent.

Willesden.—The sample inquiry in Willesden covers 167 families, 22 of which are without an earning male householder. The 167 families comprise 457 earners:

145 male householders
135 sons [1]
150 daughters [1]
27 wives and widows.

38% of all earners work locally or cycle; the inhabitants of Willesden are in this respect worse placed than those of Stepney and Tottenham, but better than those of East Ham. The expenses per head of fare-paying persons in Willesden (2*s.* 9½*d.* weekly) are, however, the lowest except for Stepney.

Willesden offers more opportunities for work to men (largely as railway employees) than to women and juveniles: 43% and 46% of the fathers and adult sons respectively work locally, as against 31% of the daughters and even less (28%) of the juvenile sons. Comparatively many housewives are gainfully employed; only just over one-quarter of them work within walking distance of home. As mentioned before, however, a number of females

[1] See note (2) on p. 157.

use public means of transport over short distances which men cover by walking. Accordingly, the average fares of females are relatively low. Adult daughters pay 2*s*. 8*d*. per week as against 3*s*. 1*d*. and 2*s*. 11*d*. paid by fathers and adult sons respectively. As regards the lower rates of juveniles (2*s*. 3½*d*. and 2*s*. 1*d*.), it must be remembered that they enjoy certain reductions for season tickets and often travel by the cheap workmen's trains, even if inconvenient. Housewives are gainfully employed in commerce and industry rather than in charing; their fares amount to 2*s*. 8*d*. per head and week.

b Analysis of Statistics of the L.C.C Housing Estates

In 1937, the London County Council made a survey of the journeys to work of the tenants of the L.C.C. housing estates. Only the householders' journeys are recorded, not those of secondary earners. Although this restriction somewhat reduces the value of the data, the table showing the mileage between home and workplace and the weekly fares is instructive (p. 190).

65,600 householders in employment (83·5% of the total tenants on L.C.C. housing estates) give information on the amount of their weekly fares, and 62,300 (78·2%) on the distance travelled. Two-thirds of all who make replies work within five miles of their homes, one-third at greater distances. There is a marked difference between the L.C.C. tenants on cottage estates and in block dwellings. Of the former, 42·4% live more than five miles from their workplace, of the latter, only 4·7%. This reflects the different location of the two types of building: cottage estates, requiring much larger sites, are situated on the fringe of the built-up area; many of their inhabitants travel to work in Central London. Tenement blocks, on the other hand, have, as a rule, been erected in inner London, many of them replacing slums; the people accommodated in them, at higher density, live nearer their workplaces.

The incidence of fares shows a corresponding difference. Over one-quarter of the flat-dwellers, but only one-eighth of the cottage-dwellers, work locally. Nearly two-thirds of the latter spend more than 3*s*. per week on the journey to work, as against under a quarter of those living in block dwellings. Every fifth cottager, but only every 33rd flat-dweller pays over 6*s*.

For the Becontree estate, the L.C.C. made a survey distinguishing the workplaces of primary and of secondary earners.[1] The primary earners number 22,500, the other earning members of the family 14,500. As the following table shows, the proportion of those who work locally [2] is approximately the same among primary and secondary earners, although somewhat higher among the latter. About three-fifths of both groups travel to London to work. Becontree is situated about ten miles from Central London.

WORKPLACES OF THE RESIDENTS OF BECONTREE HOUSING ESTATE

Type of earner	Total No. (=100)	Persons with workplaces in			
		London	Districts in which estate is situated	Districts between estate and London	Other areas
		Percentages			
Principal earners .	22,505	61·1	25·6	9·6	3·7
Secondary earners .	14,493	58·2	31·5	5·9	4·4
Total . . .	36,998	60·0	27·9	8·1	4·0

As the number of secondary earners is almost two-thirds that of the householders and as they have daily journeys of nearly the same length, the aggregate travelling expenses incumbent on the family must be put at about half as much again as the fares of the householder alone.[3] According to the table covering all L.C.C. housing estates, 29% of the Becontree tenants spend over 6*s*. weekly on the journey to work. The transport costs of these families, provided that they have the average number of earners, are, therefore, over 9*s*. per week.

[1] Barlow Commission, *Evidence*, p. 384.

[2] Namely in the districts in which the estate is situated: the boroughs of Barking and Ilford and the Urban District of Dagenham.

[3] This computation allows for a somewhat lower amount of fares paid by juveniles.

STATISTICAL TABLES

SYNOPSIS OF TRANSPORT CONDITIONS AFFECTING THE EMPLOYEES OF 36 MEMBER FIRMS OF THE INDUSTRIAL WELFARE SOCIETY, 1936

Firms (names omitted) and industry	*Town* or county	Total employees (=100)	Percentages (column 3 = 100)							
			Females	Main method of transport			Duration of journeys in hours			Going home for lunch *
				Walk	Bicycle	Train, bus, Tram	Under ½	½ to under 1	1 and over	
1	2	3	4	5	6	7	8	9	10	11
1. Wool Combing	Yorks	344	25·6	62·5	9·3	28·2	71·2	28·8 (9 and 10)		12–15 **
2. Electrical Engineers	*Liverpool*	4,223	45·4	22·5	20·1 1·4†	56·0	91·4	5·9	2·7	65
3. Carton Manufacture	*Birmingham*	370	62·2	10·8	37·8	51·4	85	15	—	49
4. Belting Manufacture	Yorks	325	38·4	51·7	12·3‡	36·0	93	6·7	0·3	55
5. Shoe Machinery Manufacture	*Leicester*	3,000	—	50	30	20	70	25	5	90
6. Paper Manufacture	Staffs	427	18·7	—	*ca.* 10	—	57·1	38·4	4·5	25
7. Clothing Manufacture	*Leeds*	9,572	24·4	—	—	—	—	—	—	30
8. Polish Manufacture	*London*	1,004	67·6	48·8	14·2 3·5†	33·5	78	19·5	2·5	—
9. Playing Cards Manufacture	*Leeds*	498	55·4	—	" quite a few "	—	—	—	—	50
10. Stores	*Liverpool*	over 1,000	—	—	" only small percentage of the males "	—	Most live within 2*d.* ride from Town Centre			—
11. Soap Manufacture	Lancs	2,402	24·7	47·4	14·6	38·0	93	6·5	0·5	54
12. Aircraft Manufacture	*Home Counties*	1,746	11·3	23·5	25·3 11·8§	39·4	58	17·7	24·3	46
13. Stationery Manufacture	Herts	4,565	49·5	20·5	—	—	68·0	31·9	0·1	50
14. Rubber Manufacture	*Birmingham*	6,100	32·9	—	35·7	—	—	—	—	40

15. Office Staff of Cement Works	Wilmington, *Hull*	214	38·3	4·6	{57·9 6·4†	31·1	—	—	—	85
16. Electrical Engineers	Beds	1,000	—	20	40–50	30–40	—	—	—	—
17. Textile Manufacture	Yorks	1,300	—	majority	11·5	—	99	1	—	69
18. Wool Combing	Staffs	315	52·4	62·8	5·2	32·0	—	—	—	56
19. Department Stores	*London*	5,935	47·4	—	7·3	—	—	—	—	less than 5
20. Soap Manufacture	Trafford Pk. *Manchester*	265	38·8	1·5	43·8‡	54·7	11·3	79·3	9·4	2
21. Ball Bearing Manufacture	Essex	3,543	36·6	51·7	42·3	6·5	—	—	—	80
22. Electrical Engineers	Middlesex	1,265	32·2	—	—	30	—	—	—	24
23. Brush Manufacture	*Bristol*	270	34·4	11·1	21·1	67·8	89·3	10·0	0·7	26
24. Photograph. Manufacture	Middlesex	2,496	29·0	—	—	—	—	—	—	80
25. Engineering	*Gloucester*	2,200	4·1	48·6	29·6	21·8	60	30	10	45
26. Engineering	Surrey	804	31·4	under 10	42·3	over 50	—	—	—	20
27. Textile Manufacture	*Leeds*	389	6·8	69·9	3·1	27·0	94	6	—	23
28. Starch Manufacture	*Hull*	2,158	55·8	23·2	31·0	45·8	97	2·7	0·3	74
29. Engineering	*Manchester*	1,600	44·3	15·0	27·0	58·0	30	60	10	47
30. Woollen Clothing	*Leicester*	—	66·6	—	"vast number"	—	—	—	—	75
31. Cleaners and Dyers	*London*	782	65·7	21·2 ‖	13·4	53·5	33·1‖	39·0	16·0	20
32. Chemical Works	Lancs	199	1·5	16·5	60·0	23·5	87	11	2	28
33. Cable Works	*London*	4,124	42·7	15·1	12·8	72·1	30	60	10	15–20
34. Shoe Manufacture	*Leicester*	529	30·0	71·5	18·1	10·0	98	2	—	94
35. Co-operative	Lancs	—	—	—	25	—	—	—	—	80
36. Motor Accessories Manuf.	*Birmingham*	1,101	30·3	28	36	36	28	72	—	27

** Of the day-shift.

* Probably including some who take lunch in cafes, etc., near the works.

† Motor cycle and private car. ‡ Including car. § 5·3 motor cycle, 6·5 private car.

‖ Plus 11·9 without answer (probably under ½ hour).

Carreras Ltd., London, N.W.1

Table 1. *Postal Districts in which Employees live, Nov.* 1936

See Chart, p. 202.

Districts Number	Name	Number of employees
East		
1.	Shoreditch, Stepney	46
2.	Bethnal Green	46
3.	Bow	69
4.	Chingford	10
5.	Clapton	14
6.	East Ham	58
7.	Forest Gate	22
8.	Hackney	21
9.	Homerton	14
10.	Leyton	27
11.	Leytonstone	21
12.	Manor Park	15
13.	Plaistow	102
14.	Poplar	72
15.	Stratford	56
16.	Canning Town, etc.	217
17.	Walthamstow	57
		867
East Central		
1.	City Rd., Old St., etc.	88
2.	Tower Hill	1
		89
West Central		
1.	Grays Inn Rd., etc.	37
2.	Long Acre, etc.	2
		39

Districts Number	Name	Number of employees
North		
1.	Islington, etc.	357
2.	E. Finchley	5
3.	Finchley	5
4.	Finsbury Park	41
5.	Highbury	19
6.	Highgate	3
7.	Holloway	86
8.	Hornsey	19
9.	Lr. Edmonton	66
10.	Muswell Hill	11
11.	New Southgate	6
12.	North Finchley	3
13.	Palmers Green	13
14.	Southgate	3
15.	South Tottenham	24
16.	Stoke Newington	21
17.	Tottenham	33
18.	Upper Edmonton	6
19.	Upper Holloway	25
20.	Whetstone	4
21.	Winchmore Hill	4
22.	Wood Green	35
		789

Table 1—*continued.*

Number	Districts Name	Number of employees
North-west		
1.	Camden Town, Euston, etc.	142
2.	Cricklewood	3
3.	Hampstead	10
4.	Hendon	3
5.	Kentish Town	34
6.	Kilburn	9
7.	Mill Hill	2
8.	St. John's Wood	3
9.	Colindale, etc.	10
10.	Willesden	6
11.	Golders Green	1
		223
South-west		
1.	Westminster, Pimlico	7
2.	Brixton	3
3.	Chelsea	3
4.	Clapham	—
6.	Fulham	3
8.	S. Lambeth	4
9.	Stockwell	6
11.	Battersea	5
12.	Balham	8
13.	Barnes	—
14.	Mortlake	3
16.	Streatham	5
17.	Tooting	11
18.	Wandsworth	4
19.	Wimbledon	4
20.	W. Wimbledon	4
		70

Number	Districts Name	Number of employees
South-east		
1.	Bermondsey Borough	22
2.	Abbey Wood	1
3.	Blackheath	2
4.	Brockley	3
5.	Camberwell	15
6.	Catford	8
7.	Charlton	1
8.	Deptford	10
9.	Eltham	1
10.	Greenwich	4
11.	Kennington	5
14.	New Cross	9
15.	Peckham	15
16.	Rotherhithe	17
17.	Walworth	21
18.	Woolwich	2
20.	Anerley	3
22.	E. Dulwich	4
23.	Forest Hill	4
24.	Herne Hill	3
25.	S. Norwood	1
26.	Sydenham	1
27.	W. Norwood	3
		155
West		
1.	Marylebone, Oxford St.	9
2.	Paddington	4
3.	Acton	3
4.	Chiswick	2
5.	Ealing	1
6.	Hammersmith	1
7.	Hanwell	—
8.	Kensington	—
9.	Maida Vale	3
10.	S. Kensington	7
11.	Notting Hill	5
12.	Shepherds Bush	9
14.	W. Kensington	3
		47
Outside London Postal Area		
	Essex	71
	Herts	8
	Kent	49
	Middlesex	143
	Surrey	34
		305

Total employees: 2,584.

Table 2. Residential Distribution, according to Time of Entering the Service of the Firm

Postal Districts in which employees live	Residential distribution * of					
	Total employees, Nov., 1935		New entrants during year		Total employees, Oct., 1936	
	No.	%	No.	%	No.	%
North-west	193	7·8	56	14·1	223	8·6
North	753	30·5	148	37·2	789	30·5
East	933	37·7	105	26·4	867	33·5
South-east	141	5·7	16	4·0	155	6·0
East Central	83	3·4	15	3·8	89	3·5
South-west	69	2·8	11	2·7	70	2·7
West	45	1·8	11	2·7	47	1·8
West Central	27	1·1	6	1·5	39	1·5
Outside London Area	228	9·2	30	7·6	305	11·9
Total	2,472	100·0	398	100·0	2,584	100·0

* Inconsistencies between the three columns are explained in the text, p. 134.

The following tables cover only the 834 employees who completed the questionnaire.

Table 3. Duration of the Journey

a. Duration according to Time of leaving Home

	Employees leaving home . . . minutes before start of work							
	Under 30	30	31–45	46–59	60	61–75	Over 75	Total
Number . .	56	86	164	82	168	185	93	834
%	17·0		19·7	30·0		33·3		100

b. Duration in Minutes

	Employees with journeys of . . . minutes							
	Under 30	30	31–45	46–59	60	61–75	Over 75	Total
Number . .	175	112	166	91	211	69	10	834
%	34·4		19·9	36·2		9·5		100

Table 4. Means of Transport used for Journeys of various Lengths

Means of transport	Employees leaving home at											
	7.30 or later		7.15 to 7.29		7.0 to 7.14		6.45 to 6.59		Before 6.45		Total	
	No.	%	No.	%	No.	%	No.	%	No.	%	No.	%
Walk . .	51	35·9	3	1·9	1*	0·4	—	—	—	—	55	6·6
Bus . .	30	21·1	23	14·0	10	4·0	2	1·0	—	—	65	7·8
Tram. .	38	26·8	63	38·4	39	15·6	33	17·9	4	4·4	177	21·2
Train. .	8	5·6	32	19·5	111	44·4	90	48·7	5	5·4	246	29·5
Combinat. of publ. convey.	6	4·2	34	20·7	74	29·6	52	28·1	83	89·2	249	29·9
Unascertained .	9	6·4	9	5·5	15	6·0	8	4·3	1	1·0	42	5·0
Total No.	142	100	164	100	250	100	185	100	93	100	834	100

* Bicycle

Table 5. *Changing of Vehicles*

Time of leaving home	Employees who change trains, etc., . . . times during each journey										Total users of publ. trans-port * = 100	No. of changes per girl
	0		1		2		3		Unascer-tained			
	No.	%	No.	%	No.	%	No.	%	No.	%		
7.30 or later	73	*80·2*	15	16·5	2	2·2	—	—	1	1·1	91	0·21
7.15–29 .	81	*49·7*	65	39·9	12	7·4	—	—	5	3·0	163	0·56
7.0–14 .	66	26·7	121	*49·0*	49	19·8	7	2·8	4	1·7	247	0·99
6.45–59 .	35	18·9	96	*52·4*	46	24·4	5	2·7	3	1·6	185	1·11
Before 6.45	8	8·6	41	44·1	36	38·7	8	8·6	—	—	93	1·48
						47·3						
Total No.	263	33·8	338	43·5	145	18·5	20	2·5	13	1·7	779	0·90

* Including 29 who are listed as unascertained in Table 4.

Table 6. *The Walk from Home to Stop or Station*

Time of leaving home	Users of public means of transport with . . . minutes' walk from home to stop or station						
	Total No.	Under 5	5–	10–	15–	20 and over	Blanks
		Percentages					
7.30 or later .	91	37·4	46·1	12·1	—	—	4·4
7.15–7.29 . .	161	21·1	47·8	26·7	2·5	—	1·8
7.0–7.14 . .	249	16·0	35·5	32·9	9·6	3·2	2·8
6.45–6.59 . .	185	13·0	35·7	28·1	19·5	3·2	0·5
Before 6.45 . .	93	14·0	26·9	37·6	16·1	5·4	—
Total . .	779	18·6	38·3	28·7	10·1	2·4	1·9
		56·9		41·2			

Table 7. Fares

a. General

Time of leaving home	Total employees	Employees who pay fares of ... *d.* per day: Nil (walk)	Under 6*d.*	6*d.*–	9*d.*–	1*s.*–	1*s.* 3*d.* and over	Blanks
				Numbers				
7.30 or later	142	51	85	6	—	—	—	—
7.15–29	164	3	136	20	2	—	—	3
7.0–14	250	1	63	129	48	9	—	—
6.45–59	185	—	7	82	66	27	2	1
Before 6.45	93	—	1	27	24	32	9	—
Total	834	55	202	204	140	68	11	4
				Percentages				
7.30 or later	100	41·8	54·7	3·5	—	—	—	—
7.15–29	100	1·8	83·0	12·2	1·2	—	—	1·8
7.0–14	100	0·4	25·2	51·6	19·2	3·6	—	—
6.45–59	100	—	3·8	44·5	35·6	14·6	1·1	0·6
Before 6.45	100	—	1·1	29·0	25·8	34·4	9·7	—
Total	100	6·6	35·1	31·7	16·8	8·1	1·3	0·4
Fare-payers (excl. walkers) = 100	779	—	37·5	34·0	18·0	8·7	1·4	0·4

Table 7—continued.

b. Workmen's Tickets

	Fare-payers			
	Total	Using	Not using	sometimes using
		Workmen's tickets		
Number	779	591	136	52
%	100	75·8	17·5	6·7

	Fare-payers not using workmen's tickets spent ... per day					
	Total	Under 6*d.*	6*d.*–	9*d.*–	1*s.* and over	Blanks
Number	136	116	16	2	1	1
%	100	85·2	11·8	1·4	0·8	0·8

Table 8. Duration of the Journey of Various Age-groups

Age-groups	Total employees		Employees leaving home at									
			7.30 or later		7.15–29		7.0–14		6.45–59		Before 6.45	
	No.	%	No.	%*	No.	%*	No.	%*	No.	%*	No.	%*
Under 20	257	30·8	70	27·2	58	22·6	59	22·9	39	15·2	31	12·1
20–24	229	35·9	47	15·7	52	17·4	91	30·4	73	24·4	36	12·1
25–29	218	26·1	22	10·1	45	20·7	72	33·0	62	28·4	17	7·8
30 and over	46	5·5	1	2·2	6	13·0	21	45·7	10	21·7	8	17·4
Blanks	14	1·7	2	—	3	—	7	—	1	—	1	—
Total	834	100	142	17·0	164	19·6	250	30·0	185	22·2	93	11·2

* % of their age-group ; first column = 100.

Table 9. *Duration of the Journey and Absenteeism due to Illness during the three years* 1933, 1934, 1935

Time of leaving home	587 girls *			581 girls * †		
	Number of girls	Absenteeism: Aggregate days	Absenteeism: Days per girl per year	Number of girls	Absenteeism: Aggregate days	Absenteeism: Days per girl per year
7.30 or later	81	1,462	6·02	80	1,307	5·45
7.15–29 . .	111	1,872	5·62 } 6·06	110	1,680	5·09
7.0–14 . .	185	3,548	6·39 }	185	3,548	6·39
6.45–59 . .	144	3,009	6·97 } 6·92	141	2,554	6·04
Before 6.45 .	66	1,349	6·81 }	65	1,213	6·22
Total .	587	11,240	6·38	581	10,302	5·91

* Employees who had stayed with the firm for three years, excluding cases of tuberculosis.

† Excluding illness of more than three months during one calendar year.

ACHILLE SERRE LTD., Walthamstow, London, E.17

Table 1. Employees, classified by Sex and Type of Work, January, 1936

Employees	Office No.	Office %	Works No.	Works %	Total No.	Total %
Male	8	12·0	260	36·4	268	34·3
Female	59	88·0	455	63·6	514	65·7
Total	67	100·0	715	100·0	782	100·0

Table 2. Postal Districts in which Employees live

Postal Sub-districts	Employees living in each district					
	Total No.	Total %	Male No.	Male %	Female No.	Female %
E.9*	119	17·2	44	20·6	75	15·6
E.17†	269	38·9	68	31·9	201	42·0
Other eastern Sub-districts	248	35·8	69	32·4	179	37·4
Total east London district	636	91·9	181	84·9	455	95·0
Essex	30	4·3	17	8·0	13	2·7
Other districts	26	3·8	15	7·1	11	2·3
Total reporting	692	100·0	213	100·0	479	100·0
Unascertained	90	—	55	—	35	—
Total employees	782	—	268	—	514	—

* District in which the works was until 1929.
† District in which the works has been since 1929.

Table 3. Duration of the Journeys

Duration in minutes	Employees						
	Male		Female		Total		
	No.	%	No.	%	No.	%	
Less than 15	33	15·4	61	12·7	94	13·6	39·0
15–29	56	26·2	120	25·1	176	25·4	
30–59	94	43·9	211	44·2	305	44·1	
60	22	10·3	76	15·9	99	14·3	16·9
Over 60	9	4·2	10	2·1	18	2·6	
Total reporting	214	100·0	478	100·0	692	100·0	

Table 4. Means of Transport

Method of transport	Employees					
	Male		Female		Total	
	No.	%	No.	%	No.	%
Walk from door to door	34	15·9	132	27·6	166	24·0
Bicycle	86	40·2	16	3·4	102	14·7
Walking or cycling		56·1		31·0		38·7
Bus	1	0·5	3	0·6	4	0·6
Tram	44	20·5	142	29·8	186	26·8
Train	8	3·7	18	3·8	26	3·8
Using either bus or tram or train		24·7		34·2		31·2
Bus and tram	31	14·5	80	16·7	111	16·0
Bus and train	1	0·5	6	1·2	7	1·0
Tram and train	6	2·8	67	14·0	73	10·6
Bus, tram and train	3	1·4	13	2·7	16	2·3
Using two or three means of transport		19·2		34·6		29·9
Unascertained	—	—	1	0·2	1	0·2
Total	214	100·0	478	100·0	692	100·0

Table 5: *Comparative Length of Journeys by Bicycle and by Public Means of Transport*

a. Alternative Statements by Sample Travellers

Employee No.	Home address	Minutes needed if travelling by	
		Bus, tram, train	Bicycle
553	Chingford, E.4	45	40
554	Clapton Park, E.5 . . .	50	40
1,739	Leytonstone, E.11 . . .	45	30
2,215	Tottenham, N.17	60	45
2,291	Clapton, E.15	45	30

b. Journeys of Several Residents of the same Postal District, Sample: Chingford, E.4

Sex	Total	Employees travelling from Chingford by											
		Bus, tram and/or train						Bicycle					
		Total	Need ... minutes					Total	Need ... minutes				
			Under 30	30	35	40	45		15	20	30	40	45
Male . .	14	4	—	1	—	2	1	10	4	4	1	1	—
Female .	27	24	—	5	7	3	9	3	—	—	2	1	—
Total . .	41	28	—	6	7	5	10	13	4	4	3	2	—

Table 6 : *The Walk from Home to Stop or Station*

Duration of walk in minutes	Employees with ... minutes' walk from home to stop or station					
	Male		Female		Total	
	No.	%	No.	%	No.	%
Under 10	103	74·7	192	58·2	295	63·0
10–14	26	18·8	89	27·0	115	24·6
15–19	9	6·5	33	10·0	42	9·0
20 and over	—	—	16	4·8	16	3·4
Over 10 minutes	35	25·3	138	41·8	173	37·0
Total users of public transport	138	100·0	330	100·0	468	100·0

Table 7 : *Going Home for Lunch*

Sex	Employees whose journey takes ... minutes								
	Under 15			15 to 29			Total under 30 *		
	Total employees = 100	Going home for lunch No.	%	Total employees = 100	Going home for lunch No.	%	Total employees = 100	Going home for lunch No.	%
Male . .	33	25	75·8	56	5	8·9	89	30	33·7
Female .	61	58	95·0	120	68	56·7	181	126	69·6
Total .	94	83	88·2	176	73	41·4	270	156	57·7

* All cases belong to journeys of under 30 minutes.

STANDARD TELEPHONES AND CABLES LTD., New Southgate, London, N.11

Table 1 : Employees, by Sex and by Type of Work, January 1936

Employees	Staff No.	Staff %	Workers No.	Workers %	Total No.	Total %
Males	662	68·5	1727	54·7	2389	57·9
Females	304	31·5	1431	45·3	1735	42·1
Total	966	100·0	3158	100·0	4124	100·0

Table 2 : Postal Districts in which Employees Live, Hourly rated employees, April 1936

See Chart, p. 204.

Postal Districts	Number of employees Male	Female	Total
N. 1	47	17	64
2	7	28	35
3	23	27	50
4	30	49	79
5	8	8	16
6	6	6	12
7	24	27	51
8	50	58	108
9	55	71	126
10	25	34	59
11	166	171	337
12	28	51	79
13	51	24	75
14	21	30	51
15	58	83	141
16	12	5	17
17	57	131	188
18	47	60	107
19	21	12	33
20	108	65	173
21	12	8	20
22	104	191	295
	960	1,156	2,116
Barnet	93	102	195
Enfield	45	51	96
Ponders End	4	19	23

Table 2—continued.

Postal Districts	Number of employees Male	Female	Total
N. Area	1,102	1,328	2,430
N.-W. 1	14	3	17
2	19	5	24
3	8	2	10
4	17	5	22
5	27	13	40
6	7	1	8
7	8	8	16
9	41	5	46
10	19	1	20
11	3	2	5
	163	45	208
Edgware	49	6	55
Harrow	10	—	10
Kenton	8	1	9
Wembley	3	—	3
Others in N.-W.	53	20	73
	286	72	358
E. 4	13	4	17
5	5	5	10
10	6	3	9
11	3	1	4
17	33	13	46
	60	26	86
Other Eastern Districts	64	18	82
E. Area	124	44	168
W. Area	43	4	47
S.-W. Area	31	2	53
S.-E. Area	86	11	97
N.-E. Area	17	2	19
Total*	1,689	1,463	3,152

*The figures for the three tables were obtained at different dates between January and April 1936 and refer to slightly different numbers of workers.

Table 3: Duration of the Journeys and Method of Transport

Main method of transport	Employees with journeys of ... minutes				
	Under 30	30–59	60 and over	Total * No.	Total * %
Walk	500	120	—	620	15·0
Bicycle	250	250	30	530	13·0
Bus	380	600	50	1,030	25·0
Train	100	1,500	330	1,930	47·0
Total No.	1,230	2,470	410	4,110	100·0
%	30	60	10	—	100·0

* See note on p. 179.

Austin Motor Co., Ltd., Longbridge, Birmingham

Table 1: Residential Distribution of Employees, Firm's Traffic Census, June 1937

See Chart, p. 206.

Place of residence	Number of Employees Day-Shift*	Night-Shift	Staff	Total
Acocks Green, Tyseley / Hall Green, Greet	235	50	23	308
Aston	88	26	—	114
All Saints', Hockley	109	20	—	129
Balsall Heath, Cannon Hill	207	52	9	268
St. Bartholomew's	12	1	—	13
Deritend, St. Martin's	110	30	2	142
Edgbaston	190	27	14	231
Erdington	94	34	6	134
Harborne, Quinton	368	49	47	464
Handsworth	193	36	10	239
Moseley, King's Heath	418	65	59	542
King's Norton	413	58	62	533
Ladywood	168	28	1	197
Lozells	79	6	2	87
Nechells	90	23	1	114
Northfield, West Heath	2,512	323	392	3,227
Longbridge Estate	173	10	27	210
Rotton Park	11	2	—	13
Perry Bar	17	3	1	21
Saltley	153	26	3	182
Sparkhill, Billesley, Yardley Wood	264	57	12	333
Small Heath	252	69	6	327
Sparkbrook	168	38	5	211
St. Mary's, Gosta Green	6	2	—	8
Bewdley	4	1	1	6
Smethwick, Bearwood	336	67	14	417
St. Paul's	21	9	—	30
Selly Park, Selly Oak / Bournville, Bournbrook	1,046	147	99	1,310
Washwood Heath, Ward End	59	2	1	62
Yardley, South Yardley, Stechford	119	32	9	160
Alcester	23	2	2	27
Alvechurch	106	18	9	133
Barnt Green	34	3	23	60
Bartley Green	18	5	1	24
Belbroughton	88	6	1	95
Brierley Hill	55	14	—	69
Bromsgrove	1,641	120	70	1,831
Blackheath	280	28	4	312
Catshill	388	19	13	420
Clent	32	—	1	33
Coventry	2	—	2	4
Cradley Heath	266	19	1	286
Darlaston	9	2	—	11
Droitwich	72	15	3	90
Dudley	207	25	2	234
Frankley	6	—	1	7
Halesowen	563	46	15	624

* Including 511 workers on continuous shifts; 213 starting work at 6 a.m., 165 at 2 p.m., 133 at 10 p.m.

Table 1—*continued.*

Place of residence	Number of Employees Day-Shift*	Night-Shift	Staff	Total
Hopwood	19	4	2	25
Kidderminster	24	1	2	27
Netherton	87	15	1	103
Old Hill	285	27	6	318
Olton	9	2	2	13
Oldbury	31	5	2	38
Quarry Bank	106	13	—	119
Rednal	518	58	98	674
Redditch	570	125	39	734
Rubery	793	56	118	967
Romsley	50	6	1	57
Stafford	1	—	—	1
Warley	87	19	3	109
Stourbridge	296	55	7	358
Wolverhampton	369	105	5	479
West Bromwich	50	10	2	62
Stirchley	252	32	21	305
Winson Green	68	14	3	85
Tipton	38	9	—	47
Langley Green	51	3	—	54
Alum Rock, Little Bromwich	70	6	—	76
Worcester	113	14	6	133
Walsall	49	8	—	57
Shirley	44	8	7	59
Sheldon	18	4	2	24
Market Hall	32	8	—	40
Blackwell	17	2	2	21
Stourport	4	—	—	4
Earlswood	3	—	—	3
Wednesbury	22	1	—	23
Solihull	6	2	3	11
Willenhall	22	1	—	23
Wythall	—	1	—	1
Rowley Regis	34	8	—	42
Sutton Coldfield	28	12	3	43
Cannock	1	—	—	1
New Oscott	18	6	2	26
Wylde Green	3	1	—	4
Castle Bromwich	1	—	3	4
Warwick	1	—	—	1
Bilston	20	8	—	28
Knowle	—	—	1	1
Marston Green	2	1	1	4
Malvern	5	—	1	6
Oakengates	—	1	—	1
Hockley Heath	—	1	—	1
Evesham	2	—	—	2
Sedgley, Coseley	25	6	—	31
Lichfield	1	—	1	2
Stratford-on-Avon	2	—	—	2
	15,945	2,296	1,310	19,551

* Including 511 workers on continuous shifts: 213 starting work at 6 a.m., 165 at 2 p.m., 133 at 10 p.m.

Table 2: Means of Transport used by the Various Types of Employees

Method of transport *	Workers			Staff	Total employees
	Day-shift	Night-shift	Contin. shifts †		
		Numbers			
Train	1,553	2	2	58	1,615
Tram	3,878	841	107	488	5,314
Bus or coach	3,915	633	225	137	4,910
Private car	2,138	261	30	277	2,706
Motor cycle	283	71	15	15	384
Pedal cycle	2,141	281	110	98	2,630
Walking	1,526	207	22	237	1,992
Total	15,434	2,296	511	1,310	19,551
		Percentages			
Train	10·1	0·1	0·4	4·4	8·3
Tram	25·1	36·6	21·0	37·3	27·2
Bus or coach	25·4	27·6	44·0	10·5	25·1
	60·6	64·3	65·4	52·2	60·6
Private car	13·8	11·4	5·9	21·1	13·8
Motor cycle	1·8	3·1	2·9	1·1	2·0
Pedal cycle	13·9	12·2	21·5	7·5	13·4
	29·5	26·7	30·3	29·7	29·2
Walking	9·9	9·0	4·3	18·1	10·2
	100·0	100·0	100·0	100·0	100·0

* i.e. the means of transport by which the employees arrive at the works.
† 213 working from 6 a.m. to 2 p.m., 165 from 2 to 10 p.m., 133 from 10 p.m. to 6 a.m.

MORRIS MOTORS, LTD., Cowley, Oxford

Residential Distribution of Employees, Firm's Census, October 1935

Cowley	1,309
Headington	717
Hinksey	30
Iffley	68
Summertown	62
Wolvercote	19
Other parts of Oxford	1,880
City of Oxford	4,086

Abingdon	177	Deddington	3
Appleford	1	Denton	1
Appleton	2	Didcot	8
Aylesbury	6	Dorchester	16
Bablock Hythe	1	Drayton	22
Baldon	3	Dry Sandford	4
Bampton	4	Ducklington	2
Banbury	54	East Hanney	1
Beckley	13	East Hendred	1
Begbroke	5	Elsfield	3
Benson	12	Enstone	2
Berrick Salome	2	Ewelme	4
Bicester	8	Eynsham	23
Bladon	16	Faringdon	3
Bletchington	1	Finstock	2
Bletchley	8	Forest Hill	12
Boars Hill	4	Garsington	92
Botley	20	Goring	1
Brackley	1	Great Haseley	2
Brill	2	Great Milton	16
Brize Norton	1	Haddenham	1
Brookhampton	1	Hailey	5
Burcot	3	Handborough	8
Burford	2	Highworth	1
Cassington	1	Holton	6
Chadlington	1	Hook Norton	2
Chalgrove	8	Horspath	62
Charlbury	2	Horton cum Studley	2
Charlton on Otmoor	2	Islip	4
Chawley	1	Kennington	24
Chilton	1	Kidlington	39
Chilworth	1	Kirtlington	4
Chinnor	1	Leafield	1
Chipping Norton	4	Lechlade	1
Chiselhampton	12	Little Milton	5
Cirencester	1	Littlemore	110
Clifton Hampden	8	Littleworth	4
Coombe	2	Long Crendon	5
Cuddesdon	22	Longworth	1
Cumnor	10	Marcham	5

Residential Distribution of Employees—continued.

Marsh Baldon	9	Stanton St. John	14
Marsh Gibbon	1	Stokenchurch	1
Marston	39	Stonesfield	5
Middle Ashton	1	Stow on the Wold	1
Minster Lovell	4	Sutton Courtenay	8
New Bradwell	3	Swindon	94
Newbury	2	Tetsworth	4
Newington	1	Thame	10
Newton	1	Thrupp	1
Noke	1	Tiddington	4
North Aston	1	Toot Baldon	13
Northleigh	1	Wallingford	27
North Stoke	1	Wantage	1
Nuneham	1	Warborough	7
Radley	14	Watlington	14
Ramsden	1	Wendlebury	1
Reading	81	Weston on the Green	5
Sandford	25	Wheatley	61
Shillingford	1	Witney	37
Shinfield	2	Woodstock	13
Shipton on Cherwell	1	Wootton	7
Stadhampton	41	Wytham	1
Stanford in the Vale	1		
Stanton Harcourt	3		5,586

SAMPLE INQUIRY IN FOUR LONDON BOROUGHS, 1929–30
(New Survey of London Life and Labour)
Families with two or more earners

Table 1: Families' Aggregate Fares for the Journey to Work

Boroughs	Total families	Families with aggregate weekly fares of . . . shillings: 0*	Up to 2	Over 2 to 4	Over 4 to 7	Over 7 to 10	Over 10	Earners per family (with two or more earners)
		a. All Families						
		Numbers						
East Ham . .	195	16	16	45	47	34	37	2·65
Stepney . . .	392	137	96	74	55	17	13	2·90
Tottenham . .	169	35	31	43	22	21	17	2·78
Willesden . .	167	27	29	33	40	27	11	2·70
Total . .	923	215	172	195	164	99	78	2·78
		Percentages						Fares per family
East Ham . .	100	8·2	8·2	23·1	24·1	17·4	19·0	6*s.* 8*d.*
Stepney . . .	100	35·0	24·5	18·9	14·0	4·3	3·3	2*s.* 8*d.*
Tottenham . .	100	20·7	18·4	25·4	13·0	12·4	10·1	4*s.* 2*d.*
Willesden . .	100	16·2	17·4	19·7	23·9	16·2	6·6	4*s.* 7½*d.*
Total . .	100	23·3	18·6	21·1	17·8	10·7	8·5	4*s.* 2*d.*
	Number = 100	*b. Fare-paying Families only* Percentages, col. 1 = 100)						
East Ham . .	179	—	8·9	25·1	26·3	19·0	20·7	7*s.* 3*d.*
Stepney . . .	255	—	37·6	29·0	21·6	6·7	5·1	4*s.* 1*d.*
Tottenham . .	134	—	23·1	32·1	16·4	15·7	12·7	5*s.* 3*d.*
Willesden . .	140	—	20·7	23·6	28·6	19·3	7·8	5*s.* 6*d.*
Total . .	708	—	24·3	27·5	23·2	14·0	11·0	5*s.* 5*d.*

* Walking, cycling, fifteen free railway passes.

Table 2: Incidence of Fares on various Members of the Family

Types of earners	Earners: Total	Earners: Working locally * No.	Earners: Working locally * %	Total † numbers	Users of public transport ‡ who pay weekly fares of ... shillings: 1s. and less	Over 1s. to 2s. 6d.	Over 2s. 6d. to 4s.	Over 4s. to 5s. 6d.	Over 5s. 6d.
					Percentages, (Col. 4‡ = 100)				
	1	2	3	4	5	6	7	8	
East Ham									
Fathers . . .	177	67	37·8	110	12·8	15·6	34·8	20·2	16·6
Sons, brothers .	154	43	27·9	111	6·6	21·7	40·6	18·9	12·2
Daughters, sisters, wives . . .	185	41	22·2	144	11·7	24·0	36·3	14·4	13·6
Total . .	516	151	29·3	365	10·5	20·8	37·2	17·4	14·1
Stepney									
Fathers . . .	332	220	66·3	112	27·6	38·4	26·8	1·8	5·4
Sons, brothers .	384	205	53·4	179	32·9	40·8	22·4	2·2	1·7
Daughters, sisters	368	198	53·8	170	25·9	47·0	24·1	3·0	—
Wives, widows .	52	30	57·7	22	22·7	63·7	9·1	4·5	—
Total . .	1,136	653	57·4	483	28·6	43·6	23·5	2·5	1·8
Tottenham									
Fathers . . .	140	55	39·2	85	5·9	44·7	33·0	8·2	8·2
Sons, brothers .	157	79	50·3	78	18·0	35·9	25·6	7·7	12·8
Daughters, sisters	159	81	53·8	78	16·7	35·9	21·8	11·5	14·1
Wives, widows .	15	10	66·7	5	(20)	(60)	(20)	—	—
Total . .	471	225	47·8	246	13·4	39·4	26·8	9·0	11·4
Willesden									
Fathers . . .	145	62	42·7	83	11·1	38·2	30·9	5·0	14·8
Sons, brothers .	135	59	43·7	76	17·8	41·1	27·4	5·5	8·2
Daughters, sisters	150	47	31·2	103	17·5	44·6	29·1	4·9	3·9
Wives, widows .	27	7	26·9	20	21·0	36·8	31·6	5·3	5·3
Total . .	457	175	38·3	282	15·9	41·3	29·4	5·1	8·3

* Including cycling to work: 18 persons in East Ham, one each in Stepney and Tottenham, 13 in Willesden.
† Including 15 holders of free railway passes.
‡ Excluding holders of free railway passes.

*Table 3: Weekly Fares paid by Adult and by Juvenile Earners of the Family**

Type of earners	Total four boroughs			East Ham			Stepney			Tottenham			Willesden		
	Total	Paying fares		Total	Paying fares		Total	Paying fares		Total	Paying fares		Total	Paying fares	
	No.	%	Per head s. d.	No.	%	Per head s. d.	No.	%	Per head s. d.	No.	%	Per head s. d.	No	%	Per head s. d.
Fathers	794	47·9	3 0½	177	62·2	3 8½	332	33·7	2 4	140	60·8	3 0½	145	57·3	3 1
Sons—															
Adult	614	54·5	2 11	113	73·5	3 8½	278	48·0	2 3	113	53·1	3 3½	110	54·2	2 11½
Juvenile	216	49·8	2 4½	41	68·3	3 4½	106	43·9	1 10	44	41·0	2 4	25	72·1	2 3½
Daughters—															
Adult	652	57·8	2 10½	134	77·6	3 9½	280	47·9	2 2	120	53·4	3 2	118	68·7	2 8
Juvenile	205	53·6	2 3	46	82·6	2 9	88	40·9	1 9½	39	35·9	2 4½	32	68·9	2 1
Wives and Widows	99	50·5	2 5	5	60·0	3 –	52	42·3	1 11½	15	33·3	2 9½	27	74·1	2 8
Grand total	2,580	53·2	2 9½	516	70·2	3 7½	1,136	42·6	2 2	471	52·2	3 –½	457	61·7	2 9½

* Owing to incompleteness of the statistical material, some juveniles may be included with adult sons and daughters.

Table 4: Cases of two or more Members of a Family working with the same Employer *

Boroughs	Total families with more than one earner	Cases of Common Workplace of two or more Members of a Family						
		No.	% of total families	Combinations:				
				Sons with father	Daughters with father	Two (or three) brothers	Two (or three) sisters	Other †
East Ham	195	29	14·8	6	4	5	10	4
Stepney	392	100	25·6	43	12	6	26	8
Tottenham	169	21	12·6	8	—	2	10	1
Willesden	167	28	16·8	11	—	6	7	4
Total	923	178	19·3	68	16	19	53	17

* Including own business. The figures of this table are somewhat casual. In East Ham, for instance, the workplace was often given as "docks" without indication whether two brothers, e.g., with such entry were employed at the same place. These cases are excluded from the table.

† Brother and sister; wife with husband; daughter with mother.

London County Council Housing Estates

Journeys to Work of the earning Householders by Distance and Amount of Fares. January, 1937

	Householders living in											
	Cottage estates											
	Becontree		St. Helier		Others		Total		Block dwellings		Total L.C.C. Estates	
	No.	%	No.	%	No.	%	No.	%	No.	%	No.	%
					A. Distance							
A. Total householders*	22,862	100	8,005	100	16,783	100	47,650	100	14,671	100	62,321	100
Of these worked												
Within 2 miles	6,962	30·5	1,605	20.1	4,881	29·0	13,448	28·2	9,946	67·8	23,394	37·6
Over 2 to 5 miles	6,317	27·6	1,306	16·3	6,371	38·0	13,994	29·4	4,041	27·5	18,035	28·9
Over 5 miles	9,583	41·9	5,094	63·5	5,631	33·0	20,208	42·4	684	4·7	20,892	33·5
					B. Weekly Fares							
B. Total householders*	23,924	100	8,105	100	18,050	100	50,079	100	15,570	100	65,649	100
Of these paid												
No fares †	2,169	9·1	692	8·5	3,741	20·8	6,602	13·1	4,276	27·5	10,878	16·5
1*s.* or under	2,236	9·3	392	4·9	405	2·2	3,033	6·1	1,246	8·0	4,279	6·5
Over 1*s.* to 2*s.*	1,523	6·3	413	5·1	1,745	9·6	3,681	7·4	2,986	19·2	6,667	10·2
Over 2*s.* to 3*s.*	2,065	8·7	606	7·5	2,486	13·8	5,157	10·3	3,321	21·3	8,478	12·9
Total up to 3*s.*	5,824	24·3	1,411	17·5	4,636	25·6	11,871	23·8	7,553	48·5	19,424	29·6
Over 3*s.* to 4*s.*	2,043	8·5	996	12·2	3,188	17·7	6,227	12·4	1,961	12·6	8,188	12·5
Over 4*s.* to 5*s.*	3,022	12·7	1,755	21·7	2,804	15·5	7,581	15·2	894	5·8	8,475	12·9
Over 5*s.* to 6*s.*	3,969	16·6	1,518	18·8	1,840	10·2	7,327	14·6	411	2·6	7,738	11·8
Total over 3*s.* to 6*s.*	9,034	37·8	4,269	52·7	7,832	43·4	21,135	42·2	3,266	21·0	24,401	37·2
Over 6*s.* to 7*s.*	3,254	13·6	912	11·2	1,032	5·7	5,198	10·4	239	1·5	5,437	8·3
Over 7*s.* to 8*s.*	1,903	7·9	431	5·3	438	2·5	2,772	5·5	95	0·6	2,867	4·4
Over 8*s.* to 9*s.*	827	3·5	206	2·6	147	0·8	1,180	2·4	47	0·3	1,227	1·9
Over 9*s.* to 10*s.*	480	2·0	107	1·3	118	0·6	705	1·4	44	0·3	749	1·1
Over 10*s.*	433	1·8	77	0·9	106	0·6	616	1·2	50	0·3	666	1·0
Total over 6*s.*	6,897	28·8	1,733	21·3	1,841	10·2	10,471	20·9	475	3·0	10,946	16·7

Source: L.C.C. Housing and Public Health Committee. Roneographed Summary of a census on the Council's housing estates of "Tenants' Workplaces and Travelling Expenses." See also agenda of the Committee, page 30, 16th November, 1937.

* Excluding unemployed householders and those giving no information. Returns were received from about 80% of all tenants.

† Working locally, cycling to work, using free railway passes.

CONCLUSION

It remains to summarise the position which the daily journey occupies in the organisation of modern life and thus to state in what ways travelling to work is a factor to be reckoned with in plans for physical reconstruction. Public appraisal of the different aspects and implications of daily travelling has undergone considerable changes in the course of time. In the current discussion of the matter, a pessimistic note prevails : the burden of the daily journey overshadows the essential services which it renders. Such an assessment of values is, however, of a recent date. It was preceded by a long period of optimism which may be said to have begun with George Stephenson, who looked forward to " the day . . . when it will be cheaper for a working man to travel on a railway than to walk on foot ".[1] The optimistic period lasted until well after the last war, when the development of outlying housing estates presupposed the journey to work as a matter of course in the daily routine of the prospective inhabitants.

At first, the daily journey was made subservient to the location of industry. Through the workers' travelling, a factory was no longer dependent upon a strictly local labour supply. Labour was, of course, not the only factor in production to become mobile. Workers took to travelling, just as raw materials came to be carried, and electric power to be transmitted, to the factory. The burdens of travelling were, on the whole, given little attention.

In the more recent past, when the evils of urban congestion were realised, the journey to work was used also for a new location of the homes of the people. Dormitories were established miles away from the noise, smell and smoke of the crowded urban centres. The problems involved in this housing policy, however, have not been fully understood, and consequently, the best use has not been made of the daily journey. Costs and hardships of travelling continued to be overlooked, and new difficulties have arisen, particularly a weakening of community life on the outskirts of the towns. The journey to work, sound in

[1] Samuel Smiles, *The Life of George Stephenson*, 1857, p. 198.

principle, has been allowed to outrun its usefulness so as to cause harm to individuals and to communities.

It is due to this development that general opinion has turned round to regarding the journey to work as a mere evil. But this is taking a narrow view. A social and economic system aiming at progress and liberty cannot do without the daily journey. There are certainly advantages in persons' living near their work, and for many this will be the desirable arrangement. For others, however, daily travelling between the places of residence and of work will be preferable, not least in the interest of maintaining family life. From the point of view of the community, moreover, the possibility of the earners' travelling to work helps to secure that flexibility which is needed to overcome the frictions inherent in any living society. The expenses in money, time and energy have therefore to be regarded as the price paid for the general benefits from daily travelling. The aim must be to enable the journey to fulfil its important functions, while keeping the movements within such limits as to avoid excessive costs, unreasonable hardships and damage to community life.

The earners' travelling to work gives planning a wide scope. On the one hand, the variety of possible combinations of dwelling places and workplaces is well suited to the multiple differentiation of modern society. On the other hand, the almost unlimited choice of lay-out which has become technically feasible confronts the town planner with an *embarras de choix*. Social and economic criteria must be used to guide the location of buildings and of zones in urban settlements. Unfortunately the information available is far from sufficient. The next Census should be used to shed light on the demographic aspects of the daily movements, proceeding on the lines of the 1921 Census but going into greater detail. In the social and economic fields, research is called for on such points as the costs of daily travelling as an item in national economy; the relation between fares and wages; the effects of the journey on industrial health and efficiency; last but not least, the implications of the severance of dwelling place and workplace for the organisation of community life. Can the elements of social and civic life at the workplace be strengthened?

While there is thus need for fuller investigation on many sides, some conclusions can be drawn even now.

That correlation between residential areas and workplaces must not be neglected needs no stressing to-day.

All parts of large towns and conurbations should be laid out at moderate compactness. Through open development, a town loses its shape and becomes unwieldy ; the central districts get cut off from the countryside ; thinness of the urban fabric impedes the growth of community life ; the journey to work is burdened with the addition of long walks between home and station. These heavy penalties of the twelve-per-acre standard are avoided by adoption of a not-so-low building density.

Nucleated development should be applied alike to dormitories, to industrial areas and to satellite towns of mixed residential and industrial character. In this respect, the foregoing examination confirms a view now widely held. Circumscribed urban units are more favourable than peripheral spread of the built-up area to the growth of local citizenship ; all kinds of services can be provided at less cost to the community ; provision of open space and access to the countryside are made easier ; finally, the journey to work is facilitated and cheapened by nucleated development.

In view of the importance of continuity for a vigorous community life, regard should be paid to preservation in spite of changing circumstances. There is a substantial difference in this respect between residential places and workplaces. Efficiency which is essential to the workplace is liable to require changes in the geographical distribution of industry. To set up a unit of production is a technical and economic task which can be achieved within a comparatively short time. It is not so easy to create a new residential unit ; the erection of houses provides but the shell. Considerable time must elapse before tradition and neighbourliness engender the social and cultural atmosphere of a community. It would, therefore, be a calamity if every change in the location of production were to entail the transference of the homes of a great number of people. There are, no doubt, occasions when the creation of new towns or of new dormitories is called for, but in many cases the better course is preservation, "redemption" and development of existing residential places. Town planning has thus to envisage, and to provide for, daily travelling from old living places to new workplaces. Regard for continuity of community life should also set limits to the pace and extent of a decentralisation of the population.

Adaptability is vital for any dynamic economic system. The journey to work is an essential means to provide it. This does not mean that the greatest possible number of persons should actually undertake daily travelling. The value of mobility of labour, as of every form of freedom, lies in the potential as well as in the actual use made of it. What is needed is to bring alternative workplaces within daily reach of every earner. The journey to work, being the most important instrument to that end, must therefore form an integral part of any scheme of physical reconstruction, as a means of combining freedom with planning.

INDEX

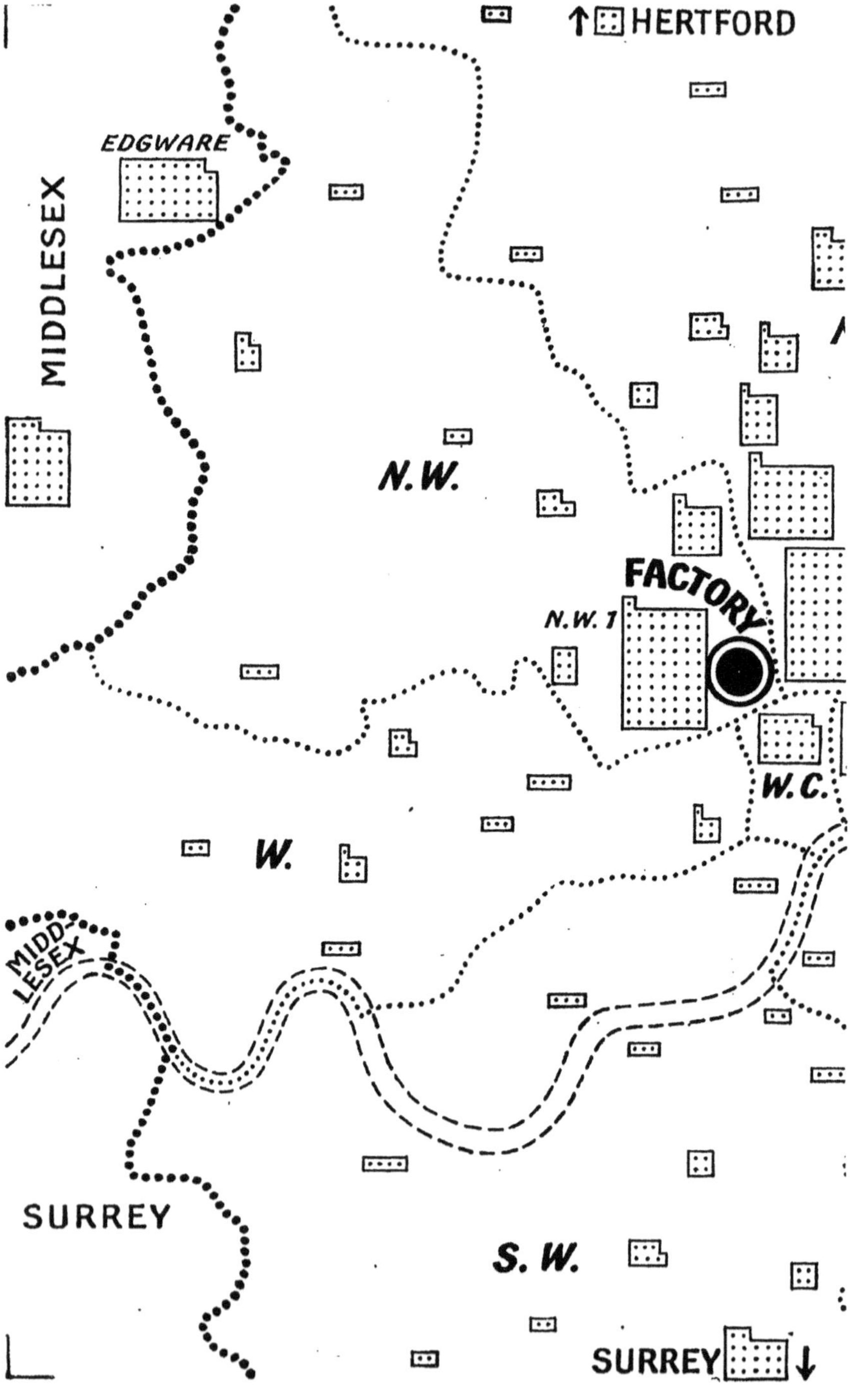
HERTFORD
EDGWARE
MIDDLESEX
N.W.
FACTORY
N.W. 1
W.C.
W.
MIDD-
LESEX
SURREY
S. W.
SURREY

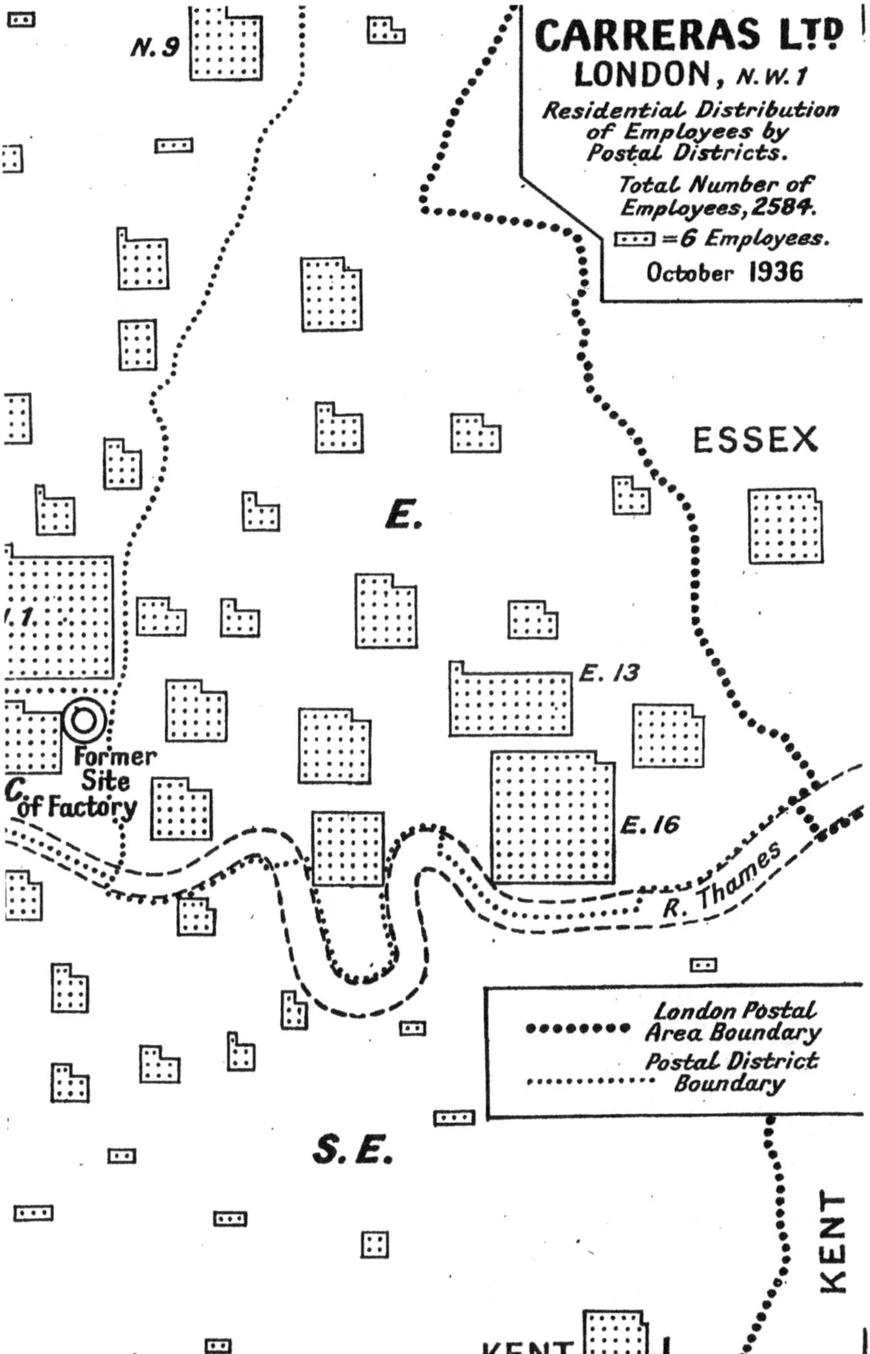
CARRERAS LTD
LONDON, N.W.1
Residential Distribution of Employees by Postal Districts.
Total Number of Employees, 2584.
=6 Employees.
October 1936
N.9
E.
E.13
E.16
ESSEX
Former Site of Factory
C.
R. Thames
London Postal Area Boundary
Postal District Boundary
S.E.
KENT
KENT

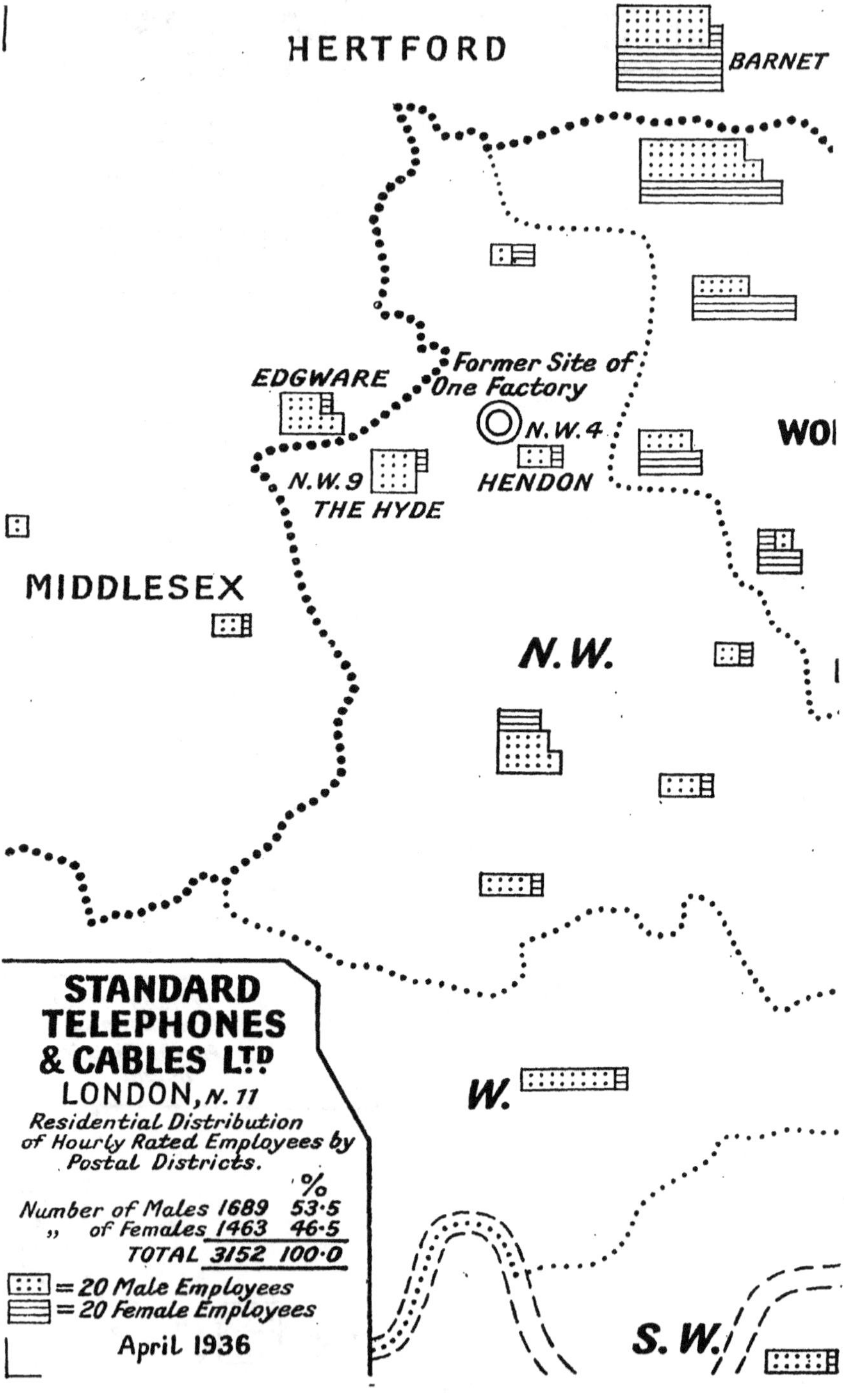
HERTFORD
BARNET
Former Site of One Factory
EDGWARE
N.W.4
N.W.9
THE HYDE
HENDON
WO
MIDDLESEX
N.W.
W.
S.W.
STANDARD TELEPHONES & CABLES LTD
LONDON, N.11
Residential Distribution of Hourly Rated Employees by Postal Districts.
%
Number of Males 1689 53·5
" of Females 1463 46·5
TOTAL 3152 100·0
= 20 Male Employees
= 20 Female Employees
April 1936

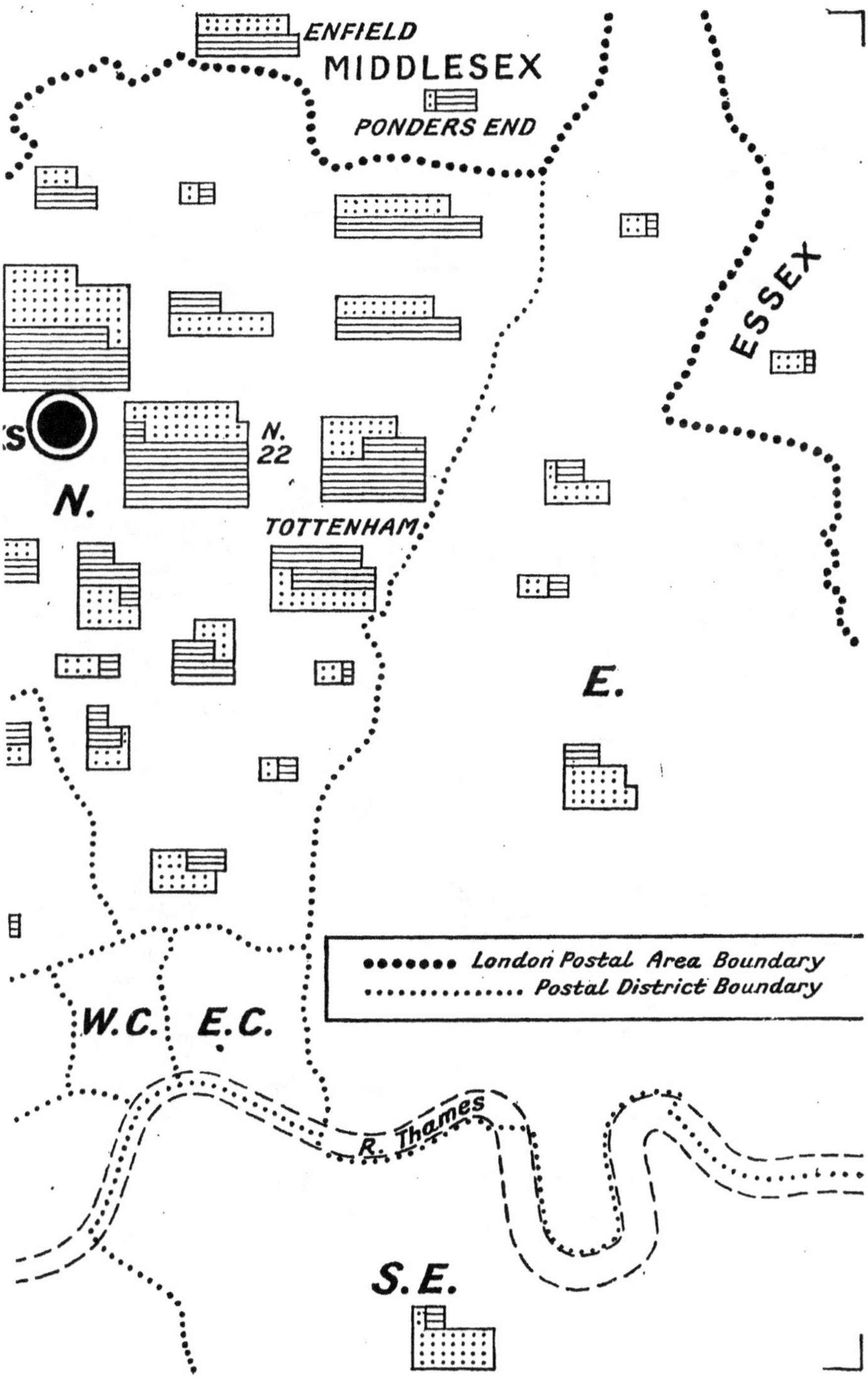
ENFIELD
MIDDLESEX
PONDERS END
ESSEX
N.
22
N.
TOTTENHAM
E.
London Postal Area Boundary
Postal District Boundary
W.C.
E.C.
R. Thames
S.E.

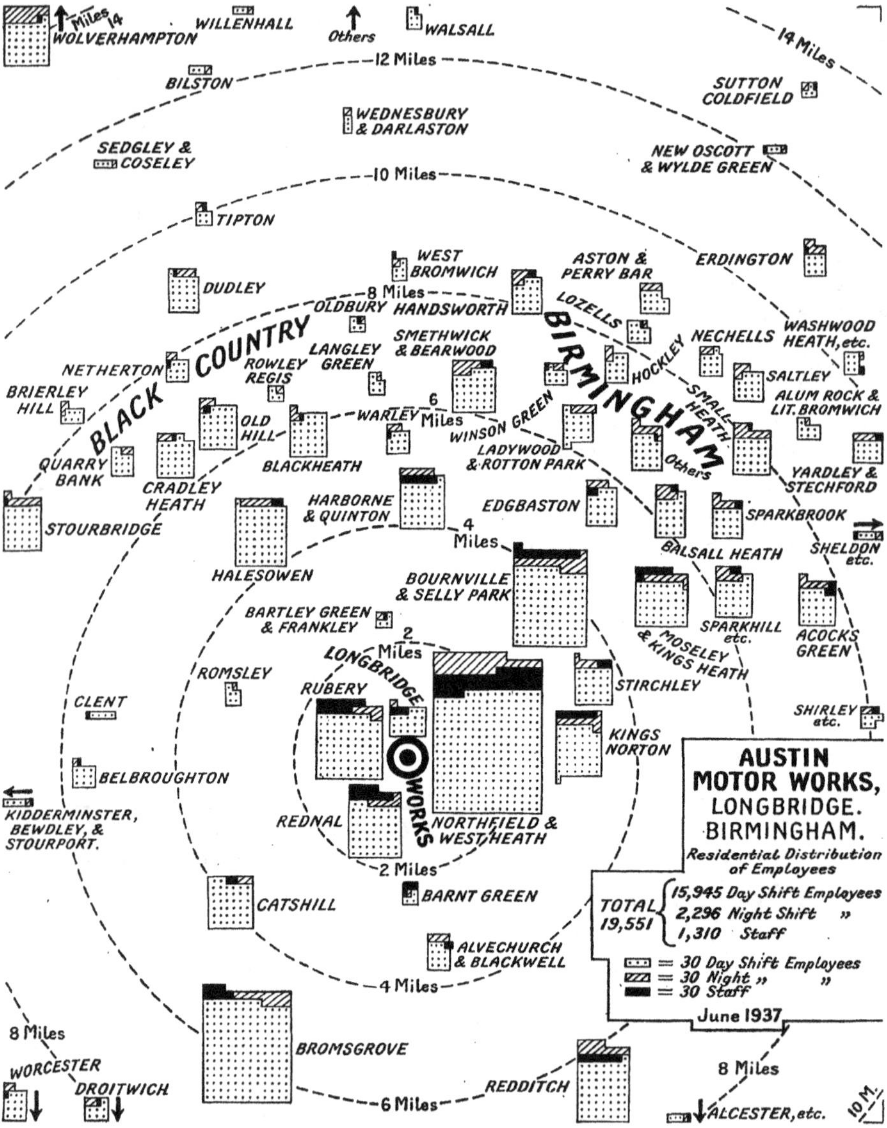
AUSTIN
MOTOR WORKS,
LONGBRIDGE.
BIRMINGHAM.
Residential Distribution of Employees
TOTAL 19,551
15,945 Day Shift Employees
2,296 Night Shift "
1,310 Staff
= 30 Day Shift Employees
= 30 Night " "
= 30 Staff
June 1937
WORKS
LONGBRIDGE
2 Miles
4 Miles
6 Miles
8 Miles
10 Miles
12 Miles
14 Miles
10 M.
BLACK COUNTRY
BIRMINGHAM
WOLVERHAMPTON
WILLENHALL
Others
WALSALL
BILSTON
WEDNESBURY & DARLASTON
SEDGLEY & COSELEY
SUTTON COLDFIELD
NEW OSCOTT & WYLDE GREEN
TIPTON
ERDINGTON
WEST BROMWICH
ASTON & PERRY BAR
DUDLEY
OLDBURY
HANDSWORTH
LOZELLS
NECHELLS
WASHWOOD HEATH, etc.
SMETHWICK & BEARWOOD
LANGLEY GREEN
ROWLEY REGIS
NETHERTON
HOCKLEY
SALTLEY
BRIERLEY HILL
SMALL HEATH
ALUM ROCK & LIT. BROMWICH
OLD HILL
WARLEY
WINSON GREEN
QUARRY BANK
BLACKHEATH
LADYWOOD & ROTTON PARK
YARDLEY & STECHFORD
CRADLEY HEATH
HARBORNE & QUINTON
EDGBASTON
SPARKBROOK
STOURBRIDGE
SHELDON etc.
BALSALL HEATH
HALESOWEN
BOURNVILLE & SELLY PARK
BARTLEY GREEN & FRANKLEY
SPARKHILL etc.
ACOCKS GREEN
MOSELEY & KINGS HEATH
ROMSLEY
RUBERY
STIRCHLEY
SHIRLEY etc.
CLENT
KINGS NORTON
BELBROUGHTON
KIDDERMINSTER, BEWDLEY, & STOURPORT.
REDNAL
NORTHFIELD & WEST HEATH
BARNT GREEN
CATSHILL
ALVECHURCH & BLACKWELL
BROMSGROVE
WORCESTER
DROITWICH
REDDITCH
ALCESTER, etc.

For Product Safety Concerns and Information please contact our EU
representative GPSR@taylorandfrancis.com
Taylor & Francis Verlag GmbH, Kaufingerstraße 24, 80331 München, Germany

www.ingramcontent.com/pod-product-compliance
Lightning Source LLC
LaVergne TN
LVHW011129300926
849160LV00008B/576
* 9 7 8 0 4 1 5 8 6 2 5 7 8 *